Brooklands Books

SHELBY COBRA

Gold Portfolio

1962-1969

Compiled by R M Clarke

ISBN 1 85520 0236

Brooklands Books

BROOKLANDS BOOKS LTD.
P.O. BOX 146, COBHAM,
SURREY, KT11 1LG. UK
sales@brooklands-books.com

www.brooklands-books.com

A-COBGP

Printed in China

Brooklands Books

MOTORING

ROAD TEST SERIES

Abarth Gold Portfolio 1950-1971
AC Ace & Aceca 1953-1983
Alfa Romeo Giulietta Gold Portfolio 1954-1965
Alfa Romeo Giulia Coupés 1963-1976
Alfa Romeo Giulia Coupés Gold Port. 1963-1976
Alfa Romeo Spider 1966-1990
Alfa Romeo Alfasud 1972-1984
Alfa Romeo Alfetta Gold Portfolio 1972-1987
Alvis Gold Portfolio 1919-1967
AMC Rambler Limited Edition Extra 1956-1969
AMX & Javelin Gold Portfolio 1968-1974
Armstrong Siddeley Gold Portfolio 1945-1960
Aston Martin Gold Portfolio 1948-1971
Aston Martin Gold Portfolio 1972-1985
Aston Martin Gold Portfolio 1985-1995
Audi Quattro Gold Portfolio 1980-1991
Audi Quattro Takes On The Competition
Austin-Healey 100 & 100/6 Gold Port. 1952-1959
Austin-Healey 3000 Ultimate Portfolio 1959-1967
Austin-Healey Sprite Gold Portfolio 1958-1971
Bentley & Rolls-Royce Portfolio 1990-2002
Berkeley Sportscars Limited Edition
BMW 6 & 8 Cyl. Cars Limited Edition 1935-1960
BMW 1600 Collection No. 1 1966-1981
BMW 2002 Gold Portfolio 1968-1976
BMW 6 Cylinder Coupés & Saloons Gold P. 1969-1976
BMW 316, 318, 320 (4 cyl.) Gold Port. 1975-1990
BMW 320, 323, 325 (6 cyl.) Gold Port. 1977-1990
BMW 3 Series Gold Portfolio 1991-1997
BMW 5 Series Gold Portfolio 1981-1987
BMW 5 Series Gold Portfolio 1988-1995
BMW 6 Series Ultimate Portfolio 1976-1989
BMW 7 Series Performance Portfolio 1977-1986
BMW 7 Series Performance Portfolio 1986-1993
BMW 8 Series Limited Edition
BMW Alpina Performance Portfolio 1967-1987
BMW Alpina Performance Portfolio 1988-1998
BMW Z3, M Coupe & M Roadster Gold Port. 1996-02
Borgward Isabella Limited Edition
Bricklin Gold Portfolio 1974-1975
Bristol Cars Portfolio
Buick Performance Portfolio 1947-1962
Buick Muscle Portfolio 1963-1973
Buick Riviera Performance Portfolio 1963-1978
Cadillac Performance Portfolio 1948-1958
Cadillac Performance Portfolio 1959-1966
Cadillac Eldorado Performance Portfolio 1967-1978
Cadillac Allante Limited Edition Extra
Checker Limited Edition
Chevrolet 1955-1957
Impala & SS Muscle Portfolio 1958-1972
Corvair Performance Portfolio 1959-1969
El Camino & SS Muscle Portfolio 1959-1987
Chevy II & Nova SS Gold Portfolio 1962-1974
Chevelle & SS Gold Portfolio 1964-1972
Caprice Limited Edition 1965-1976
Chevy Blazer 1969-1981
Camaro Muscle Portfolio 1967-1973
Camaro Performance Portfolio 1993-2000
Chevrolet Corvette Gold Portfolio 1953-1962
Chevrolet Corvette Sting Ray Gold Port. 1963-1967
Chevrolet Corvette Gold Portfolio 1968-1977
High Performance Corvettes 1983-1989
Chrysler Imperial Gold Portfolio 1951-1975
Chrysler 300 Gold Portfolio 1955-1970
Valiant 1960-1962
PT Cruiser Performance Portfolio
Citroen Traction Avant Gold Portfolio 1934-1957
Citroen 2CV Ultimate Portfolio 1948-1990
Citroen DS & ID 1955-1975
Citroen DS & ID Gold Portfolio 1955-1975
Citroen SM Limited Edition Extra 1970-1975
Shelby Cobra Gold Portfolio 1962-1969
Crosley & Crosley Specials Limited Edition
Cunningham Automobiles 1951-1955
Datsun Roadsters Performance Portfolio 1960-71
Datsun 240Z & 260Z Gold Portfolio 1970-1978
DeLorean Gold Portfolio 1977-1995
De Soto Limited Edition 1952-1960
Dodge Limited Edition 1949-1959
Dodge Dart Limited Edition Extra 1960-1976
Dodge Muscle Portfolio 1964-1971
Charger Muscle Portfolio 1966-1974
Dodge Viper Performance Portfolio 1990-1998
ERA Gold Portfolio 1934-1994
Facel Vega Limited Edition Extra 1954-1964
Ferrari Limited Edition 1947-1957
Ferrari Limited Edition 1958-1963
Ferrari Dino Limited Edition Extra 1965-1974
Ferrari Dino 308 & Mondial Gold Portfolio 1974-1985
Ferrari 328 348 Mondial Ultimate Portfolio 1986-94
Fiat 600 & 850 Gold Portfolio 1955-1972
Fiat Dino Limited Edition
Fiat Pininfarina 124 & 2000 Spider 1968-1985
Fiat X1/9 Gold Portfolio 1973-1989
Fiat Abarth Performance Portfolio 1972-1987
Ford Consul, Zephyr, Zodiac Mk. I & II 1950-1962
Ford Zephyr, Zodiac, Executive Mk. III & IV 1962-1971
Ford Cortina 1600E & GT 1967-1970
High Performance Capris Gold Portfolio 1969-1987
Capri Muscle Portfolio 1974-1987
High Performance Fiestas 1979-1991
Ford Escort RS & Mexico Limited Edition 1970-1979
High Performance Escorts Mk. I 1968-1974
High Performance Escorts Mk. II 1975-1980
High Performance Escorts 1980-1985
High Performance Escorts 1985-1990
High Perf. Sierras & Merkurs Gold Port. 1983-1990
Ford Thunderbird Performance Portfolio 1955-1957
Ford Thunderbird Performance Portfolio 1958-1963
Ford Thunderbird Performance Portfolio 1964-1976
Ford Automobiles 1949-1959
Ford Fairlane Performance Portfolio 1955-1970
Ford Ranchero Muscle Portfolio 1957-1979
Edsel Limited Edition 1957-1960
Ford Galaxie & LTD Gold Portfolio 1960-1976
Falcon Performance Portfolio 1960-1970
Ford GT40 Gold Portfolio 1964-1987
Ford Torino Performance Portfolio 1968-1974
Ford Bronco 4x4 Performance Portfolio 1966-1977
Ford Bronco 1978-1988
Shelby Mustang Ultimate Portfolio 1965-1970
Mustang Muscle Portfolio 1967-1973
High Performance Mustang IIs 1974-1978
Mustang 5.0L Muscle Portfolio 1982-1993
Mustang 5.0L Takes On The Competition
Goggomobil Limited Edition
Holden 1948-1962
Honda S500 • S600 • S800 Limited Edition 1962-1970
Honda CRX 1983-1987
Hudson Performance Portfolio 1946-1957
International Scout Gold Portfolio 1961-1980
Isetta Gold Portfolio 1953-1964
ISO & Bizzarrini Gold Portfolio 1962-1974
Jaguar and SS Gold Portfolio 1931-1951
Jaguar C-Type & D-Type Gold Portfolio 1951-1960
Jaguar XK120, 140, 150 Gold Portfolio 1948-1960
Jaguar Mk. VII, VIII, IX, X, 420 Gold Port. 1950-1970
Jaguar Mk. 1 & Mk. 2 Gold Portfolio 1955-1969
Jaguar E-Type Gold Portfolio 1961-1971
Jaguar E-Type V-12 1971-1975
Jaguar S-Type & 420 Limited Edition 1963-1968
Jaguar XJ12, XJ5.3, V12 Gold Portfolio 1972-1990
Jaguar XJ6 Series I & II Gold Portfolio 1968-1979
Jaguar XJ6 Series III Perf. Portfolio 1979-1986
Jaguar XJ6 Gold Portfolio 1986-1994
Jaguar XJS Gold Portfolio 1975-1988
Jaguar XJ-S V12 Ultimate Portfolio 1988-1996
Jaguar XK8 Limited Edition
Jeep CJ-5 & CJ-7 4x4 Perf. Portfolio 1976-1986
Jeep Wagoneer Performance Portfolio 1963-1991
Jeep J-Series Pickups 1970-1982
Jeepster & Commando Limited Edition 1967-1973
Jeep Cherokee & Comanche Pickups P. P. 1984-91
Jeep Wrangler 4x4 Performance Portfolio 1987-99
Jeep Cherokee & Grand Cherokee 4x4 P. P. 1992-98
Jensen - Healey Limited Edition 1972-1976
Lagonda Gold Portfolio 1919-1964
Lancia Aurelia & Flaminia Gold Portfolio 1950-1970
Lancia Fulvia Gold Portfolio 1963-1976
Lancia Beta Gold Portfolio 1972-1984
Lancia Stratos 1972-1985
Lancia Delta & integrale Ultimate Portfolio
Land Rover Series I, II & IIa Gold Portfolio 1948-71
Land Rover Series III 4x4 Perf. Portfolio 1971-1985
Land Rover 90 110 Defender Gold Portfolio 1983-94
Land Rover Discovery Perf. Port. 1989-2000
Fifty Years of Selling Land Rover
Lamboghini Performance Portfolio 1964-1976
Lamboghini Performance Portfolio 1977-1989
Lamboghini Gold Portfolio 1990-2004
Lincoln Gold Portfolio 1949-1960
Lincoln Continental Performance Portfolio 1961-1969
Lincoln Continental 1969-1976
Lotus Sports Racers Portfolio - covering 1951-1965
Lotus Seven Gold Portfolio 1957-1973
Lotus Elite Limited Edition 1957-1964
Lotus Elan Ultimate Portfolio 1962-1974
Lotus Elan & SE 1989-1992
Lotus Europa Gold Portfolio 1966-1975
Lotus Elite & Eclat 1974-1982
Marcos Coupés & Spyders Gold Portfolio 1960-1997
Maserati Cars Performance Portfolio 1957-1970
Maserati Cars Performance Portfolio 1971-1982
Maserati Cars Performance Portfolio 1982-1998
Matra Limited Edition 1965-1983
Mazda Miata MX-5 Performance Portfolio 1989-1997
Mazda Miata MX-5 Takes On The Competition
Mazda RX-7 Gold Portfolio 1978-1991
McLaren F1 • GTR • LM Sportscar Perf. Portfolio
Mercedes 190 & 300 SL 1954-1963
Mercedes G-Wagen 1981-1994
Mercedes S & 600 1965-1972
Mercedes S Class 1972-1979
Mercedes S Class Limited Edition Extra 1980-1991
Mercedes 230 • 250 • 280SL Gold Portfolio 1963-1971
Mercedes SLs & SLCs Gold Portfolio 1971-1989
Mercedes SLs Performance Portfolio 1989-1994
Mercedes 190 Limited Edition Extra 1983-1993
Mercedes CLK & SLK Limited Edition
Mercury Comet & Cyclone Limited Edition 1960-1970
Cougar Muscle Portfolio 1967-1973
Messerschmitt Gold Portfolio 1954-1964
MG Gold Portfolio 1929-1939
MG TA & TC Gold Portfolio 1936-1949
MG TD & TF Gold Portfolio 1949-1955
MGA & Twin Cam Gold Portfolio 1955-1962
MG Midget Gold Portfolio 1961-1979
MGB Roadsters 1962-1980
MGB MGC & V8 Gold Portfolio 1962-1980
MGB GT 1965-1980
MGC & MGB GT V8 Limited Edition
MG Y-Type & Magnette ZA/ZB Limited Edition
MGF Limited Edition
Mini Gold Portfolio 1959-1969
Mini Gold Portfolio 1969-1980
Mini Gold Portfolio 1981-1997
High Performance Minis Gold Portfolio 1960-1973
Mini Cooper Gold Portfolio 1961-1971
Starion & Conquest Performance Portfolio 1982-90
Mitsubishi 3000GT & Dodge Stealth P.P. 1990-99
Morgan Three-Wheeler Gold Portfolio 1910-1952
Morgan Plus 4 & Four 4 Gold Portfolio 1936-1967
Morgan Cars Portfolio 1968-2001
Morris Minor Collection No. 1 1948-1980
Nash Limited Edition Extra 1949-1957
Nash-Austin Metropolitan Gold Portfolio 1954-1962
Nissan Skyliner GT-R Limited Edition Extra 1989-02
NSU Ro80 Limited Edition
NSX Performance Portfolio 1989-1999
Oldsmobile Automobiles 1955-1963
Oldsmobile Muscle Portfolio 1964-1971
Cutlass & 4-4-2 Muscle Portfolio 1964-1974
Opel GT Gold Portfolio 1968-1973
Opel Manta Limited Edition 1970-1975
Packard Gold Portfolio 1946-1958
Pantera Ultimate Portfolio 1970-1995
Panther Gold Portfolio 1972-1990
Plymouth Limited Edition 1950-1960
Plymouth Fury Limited Edition Extra 1956-1976
Barracuda Muscle Portfolio 1964-1974
Plymouth Muscle Portfolio 1964-1971
Pontiac Limited Edition 1949-1960
GTO Muscle Portfolio 1964-1974
Firebird & Trans-Am Muscle Portfolio 1967-1972
Firebird & Trans-Am Muscle Portfolio 1973-1981
High Performance Firebirds 1982-1988
Firebird & Trans Am Performance Portfolio 1993-2000
Pontiac Fiero Performance Portfolio 1984-1988
Porsche 356 Gold Portfolio 1953-1965
Porsche 912 Limited Edition
Porsche 911 1965-1969
Porsche 911 1970-1972
Porsche 911 1973-1977
Porsche 911 SC & Turbo Gold Portfolio 1978-1983
Porsche 911 Carrera & Turbo Gold Port. 1984-1989
Porsche 911 Gold Portfolio 1990-1997
Porsche 911 Takes On The Competition 1990-1997
Porsche 914 Ultimate Portfolio
Porsche 924 Gold Portfolio 1975-1988
Porsche 928 Performance Portfolio 1977-1994
Porsche 928 Takes On The Competition
Porsche 944 Ultimate Portfolio
Porsche 968 Limited Edition Extra
Railton & Brough Superior Gold Portfolio 1933-1950
Range Rover Gold Portfolio 1970-1985
Range Rover Gold Portfolio 1985-1995
Range Rover Performance Portfolio 1995-2001
Range Rover Takes on the Competition
Renault Alpine Gold Portfolio 1958-1994
Riley Gold Portfolio 1924-1939
Rolls-Royce Silver Clould & Bentley S Ultimate Port.
Rolls-Royce Silver Shadow Ultimate Portfolio 1965-80
Rolls-Royce & Bentley Gold Portfolio 1980-1989
Rover P4 1949-1959
Rover 2000 & 2200 1963-1977
Studebaker Gold Portfolio 1947-1966
Studebaker Hawks & Larks 1956-1963
Avanti Limited Edition Extra 1962-1991
Subaru Impreza Turbo Limited Edition Extra 1994-2001
Sunbeam Tiger & Alpine Ultimate Port. 1959-1967
Suzuki SJ Gold Portfolio 1971-1997
Vitara, Sidekick & Geo Tracker Perf. Port. 1988-1997
Toyota Land Cruiser Gold Portfolio 1956-1987
Toyota Land Cruiser 1988-1997
Toyota Supra Performance Portfolio 1982-1998
Toyota MR2 Gold Portfolio 1984-1997
Toyota MR2 Takes On The Competition
Triumph TR2 & TR3 Gold Portfolio 1952-1961
Triumph TR4, TR5, TR250 1961-1968
Triumph TR6 Gold Portfolio 1969-1976
Triumph Herald 1959-1971
Triumph Vitesse 1962-1971
Triumph Spitfire Gold Portfolio 1962-1980
Triumph 2000, 2.5, 2500 1963-1977
Triumph GT6 Gold Portfolio 1966-1974
Triumph Stag Gold Portfolio 1970-1977
Triumph Dolomite Sprint Limited Edition
TVR Gold Portfolio 1959-1986
TVR Performance Portfolio 1986-1994
TVR Performance Portfolio 1995-2000
VW Beetle Gold Portfolio 1935-1967
VW Beetle Gold Portfolio 1968-1991
VW Karmann Ghia 1955-1982
VW Bus, Camper, Van Perf. Portfolio 1954-1967
VW Bus, Camper, Van Perf. Portfolio 1968-1979
VW Bus, Camper, Van Perf. Portfolio 1979-1991
VW Scirocco 1974-1981
Volvo PV444 & PV544 Perf. Portfolio 1945-1965
Volvo 120 Amazon Ultimate Portfolio
Volvo 1800 Gold Portfolio 1960-1973
Volvo 140 & 160 Series Gold Portfolio 1966-1975
Forty Years of Selling Volvo

CAR AND DRIVER SERIES

Car and Driver on BMW 1957-1977
Car and Driver on Corvette 1978-1982
Car and Driver on Corvette 1983-1988
C and D on Datsun Z 1600 & 2000 1966-1984
Car and Driver on Ferrari 1955-1962
Car and Driver on Ferrari 1963-1975
Car and Driver on Ferrari 1976-1983
Car and Driver on Mopar 1956-1967
Car and Driver on Mustang 1964-1973
Car and Driver on Pontiac 1961-1975
Car and Driver on Porsche 1955-1962
Car and Driver on Porsche 1963-1970
Car and Driver on Porsche 1970-1976
Car and Driver on Porsche 1977-1981
Car and Driver on Porsche 1982-1986

RACING & THE LAND SPEED RECORD

The Land Speed Record 1898-1919
The Land Speed Record 1920-1929
The Land Speed Record 1930-1939
The Land Speed Record 1940-1962
The Land Speed Record 1963-1999
The Land Speed Record 1898-1999 - Hard Bound
Can-Am Racing 1966-1969
Can-Am Racing 1970-1974
Can-Am Racing Cars 1966-1974
The Carrera Panamericana Mexico - 1950-1954
Le Mans - The Bentley & Alfa Years - 1923-1939
Le Mans - The Jaguar Years - 1949-1957
Le Mans - The Ferrari Years - 1958-1965
Le Mans - The Ford & Matra Years - 1966-1974
Le Mans - The Porsche Years - 1975-1982
Le Mans - The Porsche & Jaguar Years - 1983-91
Le Mans - The Porsche & Peugeot Years - 1992-99
Mille Miglia - The Alfa & Ferrari Years - 1927-1951
Mille Miglia - The Ferrari & Mercedes Years - 1952-57
Targa Florio - The Post War Years - 1948-1973 - H.B.
Targa Florio - The Porsche & Ferrari Years - 1955-1964
Targa Florio - The Porsche Years - 1965-1973

ROAD & TRACK 'PORTFOLIO' SERIES

Road & Track BMW M Series Portfolio 1979-2002
R & T BMW Z3, M Coupe & M Roadster Port. 1996-02
R & T Camaro & Firebird Portfolio 1993-2002
Road & Track Corvette Portfolio 1997-2002
Road & Track Dodge Viper Portfolio 1992-2002
Road & Track Ferrari V-12 Portfolio 1992-2002
Road & Track Ferrari F355 & 360 Portfolio 1995-02
R & T Jaguar XJ-S - XK8 - XKR Portfolio 1975-2003
R & T Mercedes SL - SLK - CLK Portfolio 1990-03
Road & Track MX-5 Miata Portfolio 1989-2002
Road & Track Mustang Portfolio 1994-2002
Road & Track Nissan 300ZX & 350Z Portfolio 1984-03
Road & Track Porsche 928 Portfolio 1977-1994
Road & Track Porsche 911 Portfolio 1990-1997

ROAD & TRACK 'ON' SERIES

Road & Track on Aston Martin 1962-1990
Road & Track on Audi & Auto Union 1952-1980
Road & Track on Audi & Auto Union 1980-1986
Road & Track on Austin Healey 1953-1970
Road & Track on BMW Cars 1966-1974
Road & Track on BMW Cars 1975-1978
Road & Track on BMW Cars 1979-1983
R & T on Cobra, Shelby & Ford GT40 1962-1992
Road & Track on Corvette 1953-1967
Road & Track on Corvette 1968-1982
Road & Track on Corvette 1982-1986
Road & Track on Corvette 1986-1990
Road & Track on Ferrari 1975-1981
Road & Track on Ferrari 1981-1984
Road & Track on Ferrari 1984-1988
Road & Track on Fiat Sports Cars 1968-1987
Road & Track on Jaguar 1950-1960
Road & Track on Jaguar 1961-1968
Road & Track on Jaguar 1968-1974
Road & Track on Jaguar 1974-1982
Road & Track on Jaguar 1983-1989
Road & Track on Lamborghini 1964-1985
Road & Track on Mercedes 1952-1962
Road & Track on Mercedes 1963-1970
Road & Track on Mercedes 1971-1979
Road & Track on MG Sports Cars 1949-1961
Road & Track on MG Sports Cars 1962-1980
Road & Track on Pontiac 1960-1983
Road & Track on Porsche 1951-1967
Road & Track on Porsche 1968-1971
Road & Track on Porsche 1972-1975
Road & Track on Porsche 1975-1978
Road & Track on Porsche 1979-1982
Road & Track on Porsche 1985-1988
R & T on Rolls Royce & Bentley 1950-1965
R & T on Rolls Royce & Bentley 1966-1984
Road & Track on Saab 1972-1992
R & T on Toyota Sports & GT Cars 1966-1984
R & T on Triumph Sports Cars 1967-1974
R & T on Triumph Sports Cars 1974-1982
Road & Track on Volkswagen 1951-1968
Road & Track on Volkswagen 1968-1978
Road & Track on Volkswagen 1978-1985
Road & Track on Volvo 1957-1974
Road & Track on Volvo 1977-1994

ROAD & TRACK BY AUTHOR

Road & Track - Henry Manney at Large & Abroad
Road & Track - Best of PS
Road & Track - Peter Egan "At Large"
Road & Track - Peter Egan Side Glances 1983-92
Road & Track - Peter Egan Side Glances 1992-97
Road & Track - Peter Egan Side Glances 1998-02

PRACTICAL CLASSICS SERIES

PC on Land Rover Restoration
PC on Midget/Sprite Restoration
PC on MGB Restoration
PC on Sunbeam Rapier Restoration
PC on Triumph Herald/Vitesse

HOT ROD 'ENGINE' SERIES

Chevy 265 & 283
Chevy 302 & 327
Chevy 348 & 409
Chevy 396 & 427
Chevy 454 thru 512
Chrysler Hemi
Chrysler 273, 318, 340 & 360
Chrysler 361, 383, 400, 413, 426 & 440
Ford 289, 302, Boss 302 & 351W
Ford 351C & Boss 351
Ford Big Block

RESTORATION & GUIDE SERIES

BMW 2002 - A Comprehensive Guide
BMW '02 Restoration Guide
Classic Camaro Restoration
Chevrolet High Performance Tips & Techniques
Chevy-GMC Pickup Repair
Engine Swapping Tips & Techniques
Land Rover Restoration Portfolio
Lotus Elan Restoration Guide
MG 'T' Series Restoration Guide
MGA Restoration Guide
Mustang Restoration Tips & Techniques
Practical Gas Flow
Restoring Sprites & Midgets an Enthusiast's Guide
SU Carburetters Tuning Tips & Techniques
The Great Classic Muscle Cars Compared

MILITARY VEHICLES

Complete WW2 Military Jeep Manual
Dodge WW2 Military Portfolio 1940-1945
German Military Equipment WW2
Hail To The Jeep
Combat Land Rover Portfolio No. 1
Land Rover Military Portfolio
Military & Civilian Amphibians 1940-1990
Off Road Jeeps Civilian & Military 1944-1971
US Military Vehicles 1941-1945
US Army Military Vehicles WW2-TM9-2800
VW Kubelwagen Military Portfolio 1940-1990
WW2 Allied Vehicles Military Portfolio 1939-1945
WW2 Jeep Military Portfolio 1941-1945

MOTORCYCLES

To see our range of over 60 titles visit www.brooklands-books.com

14/02Z4

Brooklands Books

CONTENTS

Brooklands Books

ACKNOWLEDGEMENTS

The car that we know as the Cobra today started life in Thames Ditton near London in 1953 as the AC Ace. In those days it was fitted with an ancient six cylinder AC engine which did not produce the performance that was needed for such a beautiful sportscar.

In the late fifties the AC engine was replaced by a 2 litre Bristol unit with which it went on to win many races on both sides of the Atlantic. Sadly the Bristol engine went out of production about 1960 and the following year AC fitted a tuned version of the British Ford Zephyr engine in the Ace.

Carroll Shelby then entered the picture. In September 1961 he wrote to AC asking if they could supply cars capable of taking a US Ford V8 engine. The Ace chassis and body had just been revamped to take the larger Zephyr engine so all that was needed to cope with the extra power was to beef-up the differential, modify the brakes and engine mounts, and fit larger wheels which were taken from AC's heavier Greyhound sedan. The outcome was that the first Cobra - sans engine—was shipped to the US in February 1962. To this Shelby quickly fitted a Ford Fairlane 260 power unit and had it ready in April for the New York Auto Show.

The story of the Cobras' development is fully covered in the pages that follow. The Cobra name was sold by Shelby to Ford in the mid-sixties and therefore some 'Cobras' that were manufactured after this time carried AC markings. The AC Autokraft Mk IV is a direct descendant of the early Cobras and for this reason is included here.

This Gold Portfolio could not have been produced without the co-operation of the original publishers and authors. Our thanks go especially to the management of *Auto Topics, Autocar, Autosport, Car, Car and Driver, Classic & Sports Car, Cobra / Mustang Guide, EVO, Hot Rod, Motor, Motor Sport, Motor Sport US, Motor Trend, Practical Classics, Road & Track, Road & Track Specials, Sports Car Graphics, Sports Car Specials, Sports Car World, Vintage Racer and World Autos* for allowing us to include their copyright road tests and other stories in this anthology.

R.M.Clarke

ROAD TEST/21-62

SHELBY AC COBRA

ROAD TEST AND TECHNICAL REPORT

At top, SCG's test crew starts out on one of the runs with fifth wheel and electric speedometer which pegged its needle at 100 well before end of quarter mile. Above and left, at full noise in third, car sticks to the road like paint.

PHOTOS BY BOB D'OLIVO

"GO AHEAD, put your boot in it," said Carrol Shelby, "you're not going to hurt it that way!"

Carrol was referring to the fact that on our first trip with his then brand new Cobra (SCG, May 1962) we had been reluctant to push it since the car was the only one around at the time.

Since then this latest addition to the really hot production car ranks has been around some. A smash hit at the New York Automobile Show, the pearlescent yellow roadster was equally popular in Dallas. After the Dallas show it was literally kidnapped and dragged off to Dearborn by the Ford people. At either show Shelby could have disposed of his entire year's production to eager Ford dealers who literally thrust deposits at him. However, with an initial planned production run of 350, it has been decided that there will be only three dealers in key areas, only one of whom, Ed Hugus of Philadelphia, has been appointed. Shelby himself will handle the Western distribution.

To get back to that instruction noted above, we did just that. The first thing we discovered was that it was literally impossible to get true readings from the car's own instrumentation during hurried getaways. Wheelspin, in spite of the limited slip differential, produced excessive speedo and tachometer readings, particularly in low gear. Consequently, for all runs up to 100 mph a fifth wheel and electric speedometer were used. Even that posed a problem since the top figure on the unit was 100 and the meter was pegged long before the end of the measured quarter mile – in fact it was exceeded before the car ran out of revs in Third gear.

When we first tried the car, as reported in May, it was fitted with the engine that is to be standard, the special series of 260 cubic inch Fairlane V-8s made by Ford for this project. It was fitted with a small four-barrel Holley carburetor and 9.2 to 1 heads. Since then it has been decided to use a slightly larger carburetor on the standard engine and, for our test, a fresh engine with this larger unit was installed, the original engine (XHP 260-1) having been well and truly thrashed both in the car and on the dynamometer. The result of the mildly increased breathing was slightly less initial torque at the bottom end, where it really isn't needed, and lots more in the middle and upper ranges.

With the 3.54 to 1 rear end gearing the standard engine

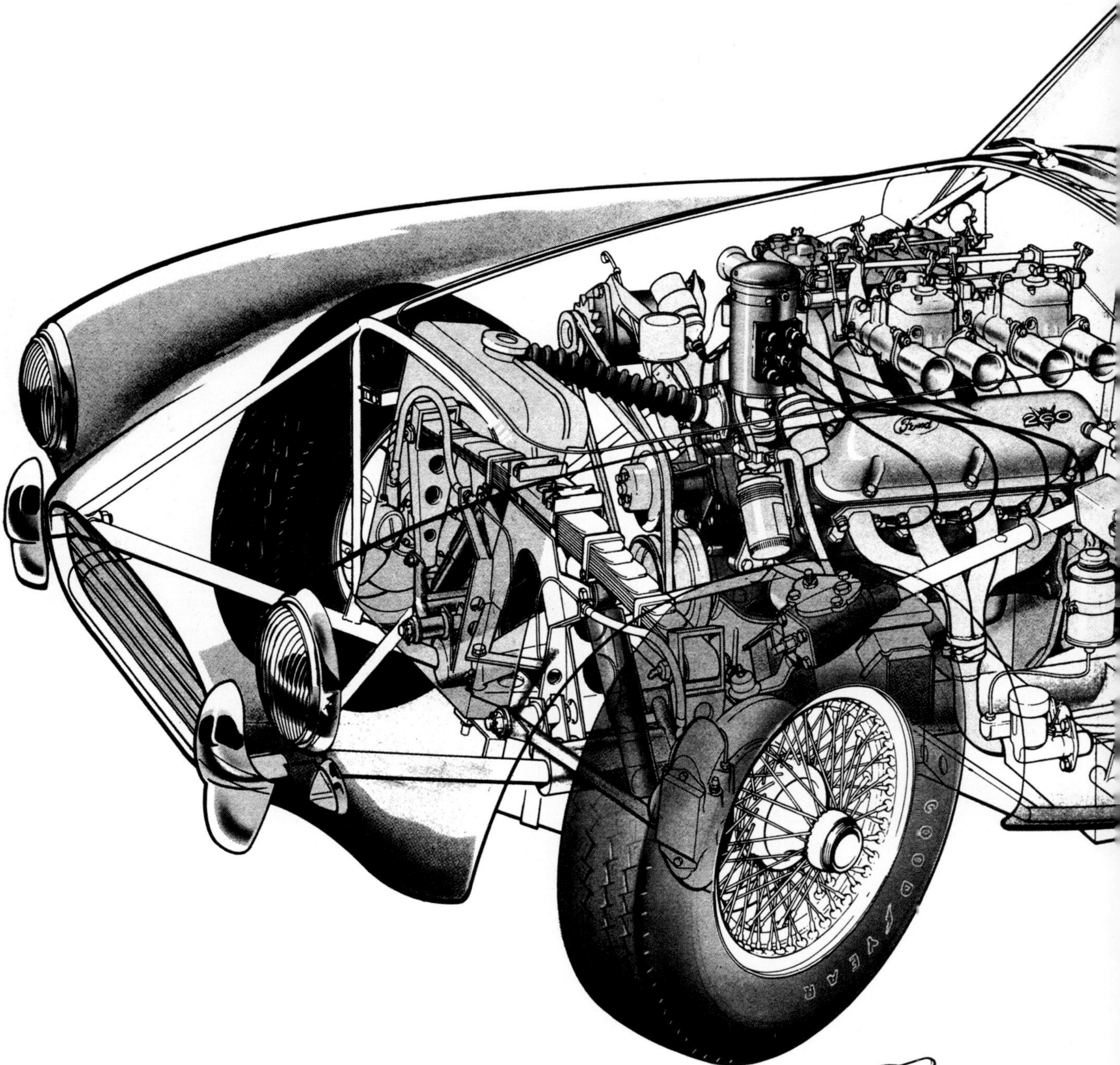
Ford
260
GOODYEAR
DARLING

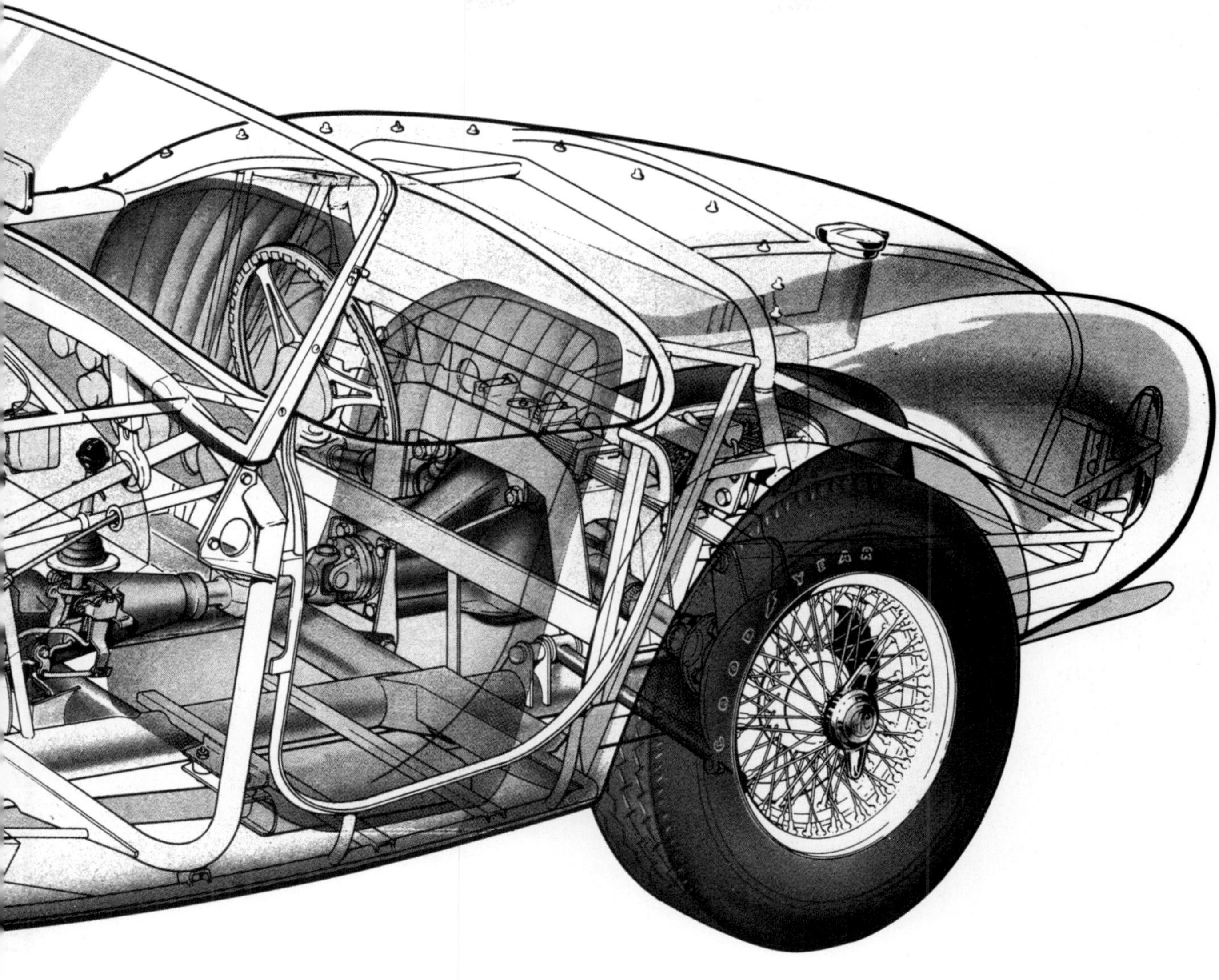

SHELBY AC COBRA

will develop 7200 rpm in all four gears which, disregarding tire growth, translates to a top speed of 151 mph, reaching this in a space of less than a mile. In this particular case tire growth was a negligible factor since the Cobra was shod with Goodyear low-profile racing tires which show almost no distortion at speed. Using a standard profile tire, with its additional half-inch of loaded radius, the top speed would probably be right around 155 mph. However, it is our feeling that in this particular case the small sacrifice at the top end was more than made up by the extra bite afforded by the wide, flat treads of the Goodyears which complemented the normally fierce side bite that is one of the Cobra's best traits.

Pushing the car hard into Riverside Raceway's Turn 9, a banked 180-degree bend that is slower than it looks, was an experience that was unique for production equipment — sort of as though someone up there had suspended one or more natural laws having to do with centrifugal force. In brief, the car sticks under conditions in which most other cars would have everything hanging out short of Monday's wash. During one photo session we brought the Cobra through a bend with the idea of popping the back end out in the middle of the turn for the photographer. A poke at the throttle at the proper moment provoked only more velocity and no change of attitude, indicating that we had far under-rated the Cobra's cornering speed. In fact, the tenacity of this vehicle is such that we would want far more time than was available (and a good roll bar) before we tried any on-the-limit barreling. As with certain small modified machines, the upper limit is such that it takes practice and a double handful of brave pills to find the point where adhesion ends and sliding begins.

About the only point on which we could find fault with the car mechanically was the gear splits in the Borg-Warner all-synchro box. Second and Third gears are nicely spaced in regard to each other but First is on the low side and the spacing between Third and Fourth is just a little wider than need be, at least with the 3.54 rear-end gearing. Dropping from top to Third for acceleration, as in highway passing, produces the acceleration but it winds out so quickly that Fourth has to be snatched sooner than would seem necessary. With the alternate 2.92 gearing, however, this fault would probably be considerably less noticeable. Too, Shelby plans an alternate choice of transmission gears for those who like their cogs closer together. Another point that might be mentioned is that this wide split factor is helpful around town in that Third is good for anything from a slow crawl on up to and above legal highway speeds, almost like having an automatic transmission in its effect.

Perhaps one of the most outstanding attributes of the Cobra, aside from the combination of docility and brutal performance, is the fact that for its type it is a very liveable car. There is plenty of leg and shoulder room in the cockpit and the seats, in true Ace fashion, are deep and comfortable with more than adequate side and shoulder support. In addition, with the standard 18-gallon fuel tank, luggage space is as generous as can be found in virtually any sports roadster and far more capacious than most. All things considered, this is one of the few cars that can conceivably be driven all during the week, then loaded with luggage, driven to the local *piste de vitesse*, raced and then be driven home again. In fact, only one or two other marques, all in the five-figure price class, come to mind in this respect.

For the benefit of those who may not have read our earlier report on the Cobra we'll repeat the description of the car. In design it basically follows the plan laid down in the Fifties by John Tojiero for the Ace and A.C. Bristol, using large-diameter tubing for the main frame rails, with bridge-structures at each end. Suspension, attached to these structures, is fully independent, with A-arms and transverse leaf springs which, in the Cobra, are fully wrapped with heavy lube impregnated fabric. The whole is topped by a tube-framed aluminum body. At first blush it looks like the famous Ace but there the resemblance ends.

First and foremost the Cobra is not merely an Ace with a Ford stuffed under the hood. That sort of thing has been tried before and has almost always been unsuccessful. Virtually everything but the frame rails has been redesigned, beefed up or replaced. The suspension arms are considerably stronger and reinforced with plating, the springs are considerably stiffer and the track has been widened four inches. Wheels used are 72-spoke Dunlops in the 15-inch size, rather than the narrower 16-inch units used before. The differential has been replaced with a much larger Salsbury unit, as fitted to the XKE Jaguar, and takes the same

(continued)

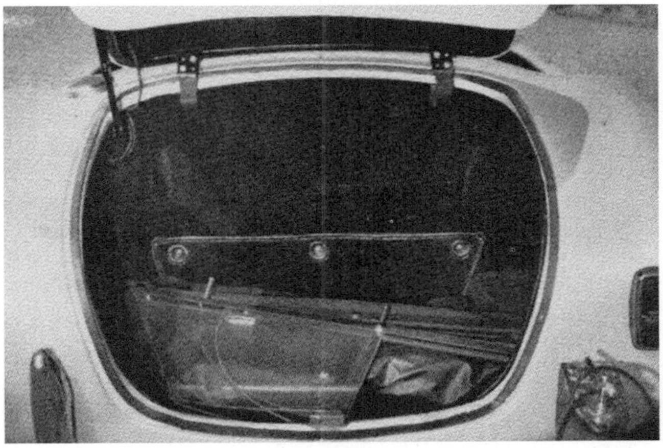

At left, runs up to 100 mph were done with fifth wheel and electric speedometer to negate tremendous wheelspin effect in lower gears. Cobra is fastest accelerating production sports car ever tested by SCG. Above, cockpit is more fully instrumented than most and trunk is large for this type car.

With only some of the options big increases in power are gained. With just the four Webers shown, engine picked up 42 bhp. Single carb form, right, with optional cam and increased compression, torque jumped to 300 lbs/ft. With all options power is over 335 hp sustained with an extremely wide usable range. For those not interested in production racing almost any combination can be purchased, up to and including 300 CI.

range of gears. The body has been widened at the wheel wells and flared to accept the wider track, and has also been given an attractive "nose job" with a smaller grille replacing the rather blunt front aspect of the older Ace. In place of the flat windshield that was used on the 2-liter is a nicely curved screen with considerably greater rearward rake; aside from attractiveness the main attribute of this change has been to reduce wind turbulence in the cockpit.

As mentioned, the cockpit is one of the most liveable of any sports roadster. Except for instrumentation changes it is virtually the same as in the Ace. Virtually every gauge that can be used is mounted on the wide leather-faced dash. The tachometer which will be used on the production version will be an 8000 rpm unit in place of the 6000 rev instrument in the test car. The speedometer will, in all likelihood, be the excellent 180 mph VDO unit as used in the Mercedes 300 SL. Driving position is of the "arms out" variety but this can be modified through means of the adjustable-for-distance wheel.

The Cobra is also one of those vehicles which can literally be tailored to the customer through an option list as long as any. For those not interested in marque or class racing, there is even a 300 cubic inch engine. For the 260-inch engine there is a range of listed options that can deliver anything up to 330 horsepower in almost any useable curve.

Carburetion options listed on the homologation papers submitted to the FIA are a case in point. There is the standard four-throat unit as on the test car, a dual quad, a triple two-throat set-up and, finally, the very hairy-looking quadruple twin-choke Weber combination on a ram-tuned manifold. In one test this was bolted on a dyno-thrash standard engine that had shown only 242 bhp. The addition of the Webers *alone*, with no other changes, produced 282 bhp, an addition of 40 horses for the one item. In another test, using the single quad carburetor, but with one of four optional cams and 11 to 1 compression ratio, the engine developed 300 lbs./ft. of torque – 31 more than the standard version. Power for the full-option engine is pegged at 335 sustained horses and the claim, judging from single-option tests, is very likely modest.

Why, if the standard version performs so creditably, is there a need for all this optional soup? Simply that the automotive world is not standing still. Chevrolet is known to have several good things in its future file for the Corvette, Jaguar has an increasing option list and, as is common knowledge, the latest Ferrari Berlinetta is little more than a Testa Rossa with a roof. While the standard version of the Cobra may well be competitive currently in SCCA racing it wouldn't long remain that way, especially in International events. Nobody is more aware of this than Shelby and he intends to see that his customers get everything they wish in order to keep up with the competition.

Naturally all this soup-kitchen business will not come free but the cost on a percentage basis is surprisingly low. As is well and painfully known, competition preparation and full use of the option list on any production car can come close to doubling the initial cost, at least if one wants to field a winning combination. In the case of the Cobra, however, the cost of a fully race-prepared version – other than the usual individual tailoring – is an additional one-sixth the base purchase price. For the FOB-Los Angeles tab of $5,995 the customer gets a going bear that will do well in club racing. The full race version is tentatively priced at $7,200, give or take a dollar, and for this the customer buys a projectile that will do honest battle with anything in the world that is classed as production.

Shelby's brochure says boldly – and impudently – "Buy it or watch it go by." He means what he says. Put differently, the Cobra is one way to go Superfast and save $10,000.

— John Christy

TEST DATA
ROAD TEST NO. 21/62

VEHICLE	Shelby AC Cobra	MODEL	Standard
PRICE (as tested)	$5995.00 P.O.E.	OPTIONS	None

ENGINE:

Type	Ford 260 V-8
Head	Cast iron, removable
Valves	OHV, pushrod/rocker
Max. bhp	260 @ 6500 rpm
Max. Torque	269 lbs. ft. @ 3600 rpm
Bore	380 in. 96.5 mm.
Stroke	2.87 in. 72.9 mm.
Displacement	260 cu. in. 4262 cc.
Compression Ratio	9.2 to 1
Induction System:	Single, four-barrel, cast manifold
Exhaust System	6-into-2 headers
Electrical System	12V distributor, ignition

CLUTCH:

Single disc, dry	
Diameter	9.5 in.
Actuation	Hydraulic

TRANSMISSION:

4-speed, full synchro	
Ratios: 1st	2.36 to 1
2nd	1.78 to 1
3rd	1.41 to 1
4th	1.0 to 1

DIFFERENTIAL:

Ratio	3.54 to 1
Drive Axles (type)	two-joint, open

STEERING:

Worm & Sector	
Turns Lock to Lock	3
Turn Circle	31 ft.

BRAKES:

Disc rear	
Drum or Disc Diameter	12 in.
Swept Area	550 sq. in.

CHASSIS:

Frame	Tubular steel
Body	Aluminum
Front Suspension	Lower A, transverse leaf upper arm, tube shocks
Rear Suspension	I.R.S., Lower A's, transverse leaf upper, tube shocks
Tire Size & Type	6.40 x 15 front, 6.70 x 15 rear Goodyear SS

WEIGHTS AND MEASURES:

Wheelbase	90 in.	Ground Clearance	7.0 in.
Front Track	51.5 in.	Curb Weight	2020 lbs.
Rear Track	52.5 in.	Test Weight	2385 lbs.
Overall Height	49.0 in.	Crankcase	10 qts.
Overall Width	61.0 in.	Cooling System	N.A. qts.
Overall Length	151.5 in.	Gas Tank	18 gals.

PERFORMANCE:

0-30	1.8 sec.	0-70	5.4 sec.
0-40	2.5 sec.	0-80	6.8 sec.
0-50	3.3 sec.	0-90	8.0 sec.
0-60	4.1 sec.	0-100	10.8 sec.
Standing ¼ mile	12.9 sec. @ 114 mph		
Top Speed (av. two-way run)	152 mph		
Fuel Consumption: Test	12 mpg	Average:	16 mpg
RPM Red-line	7200 rpm		
Speed Ranges in Gears:			
1st	0 to 41 mph	3rd	22 to 109 mph
2nd	12 to 78 mph	4th	36 to top mph
Brake Test	80 Average % G, over 10 stops		

Fade not encountered

REFERENCE FACTORS:

BHP per Cubic Inch	1.0
Lbs. per bhp	7.77
Piston Speed @ Peak rpm	3109 ft./min.
Sq. In. Swept Brake area per Lb.	0.273

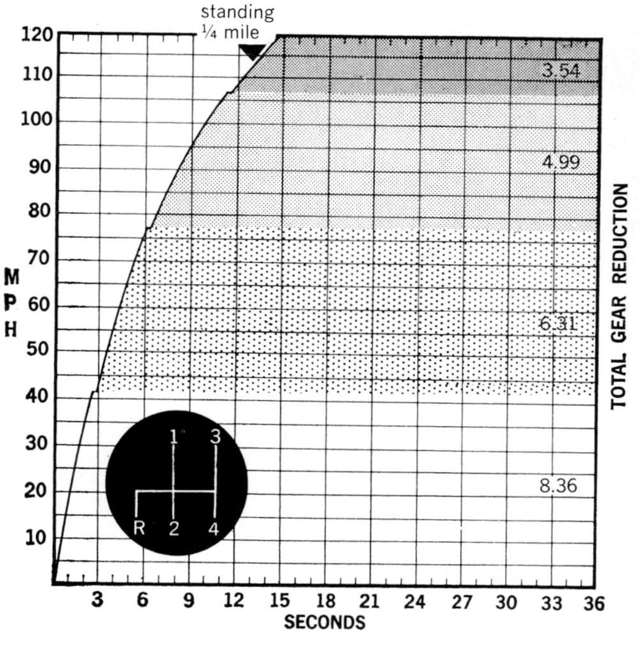

COBRA 427

JOHN BOLSTER tests the

A.C. Cobra

POWERED BY Ford

THE A.C. Cobra is at present an export-only model. It is based very closely on the A.C. Ace, which normally carries a Bristol or Ford Zephyr engine. Suitably strengthened to withstand an American Ford V8 unit of 4$\frac{3}{4}$ litres capacity, the Cobra is manufactured at the A.C. factory at Thames Ditton and shipped to the States, where car and engine meet. About 10 Cobras per week are sent over to Shelby American, Inc., who are selling them like hot cakes.

The A.C. chassis was designed by John Tojeiro, and was in fact based on the very successful Bristol-engined "Toj" raced by Cliff Davis. It has a ladder-type tubular frame, with independent suspension at both ends by transverse springs and wishbones. The Cobra has tubes of heavier gauge and the shorter Vee-type engine has permitted the insertion of extra cross-bracing. Rack and pinion steering and anti-roll torsion bars front and rear are additions to the specification, and naturally the differential and universally jointed half-shafts are heavier. The differential is of the self-locking variety and the track has been slightly increased, both front and rear.

Panelled in aluminium, the open two-seater body has a lower grille and bonnet than that which the overall height of the Bristol engine demands. The Ford unit, with its own all-synchromesh four-speed gearbox, is in fact notably compact. With the over-square dimensions of 101.6 mm. x 72.9 mm. (4,727 c.c.), it probably develops around 275 b.h.p. in touring trim and over 350 b.h.p. when tuned for racing. All this in a car weighing some 18 cwt. can only result in a simply shattering performance.

I was thus more than delighted when, during some glass-lifting exercises at the Steering Wheel, Carroll Shelby suggested that I should borrow his personal Cobra for a day. To make the most of the occasion, that day started before 5 a.m.! Carroll had remarked that the car was "set up for the street", which means that it was neither geared for an ultimate maximum speed nor fitted with the greatest available number of carburetters. It was, in fact, far more flexible

on top gear than any luxury limousine and its traffic manners were impeccable.

What can one do with a car which will out-accelerate practically anything on the road when top gear is engaged at 15 m.p.h., yet has about the best close-ratio gearbox that could be imagined? The desire to feel that "whoosh" of power from zero revs. on top is always competing with the excitement of rushing past 50 m.p.h. on bottom gear. With the final drive ratio fitted to the test car, 6,500 r.p.m. came up at once in top and I lifted my foot. This resulted in a timed speed of 136 m.p.h., but of course 150 m.p.h. could be exceeded with a suitable crown wheel and pinion in place.

The engine must be praised very highly indeed. A new lightweight Ford model, it is utterly smooth at all speeds up to 6,500 r.p.m., and would certainly go much higher if one turned a blind eye to the dial. The clutch copes happily with the mammoth torque and the four synchronized gears may be selected with the greatest ease and rapidity. The roads were damp, though not wet, when I started my performance tests, and rain was actually falling at their completion. As a result, the figures which I recorded could be beaten on a dry road. Nevertheless, to accelerate from a standstill to 50 m.p.h. in 3.8 seconds or to 80 m.p.h. in 8.2 seconds is a somewhat dramatic performance.

Very quiet mechanically, the power unit has just a trace of exhaust "beat". This noise, so typical of big V8s, is rather objectionable to most hearers and could probably be eliminated by some cross connection of the two separate exhaust branches. It would appear that V8 noises are accepted in America, but in England they are no more popular than were the single-cylinder detonations

of the 500 c.c. racers. This is a minor point, however.

The A.C. chassis has been improved by the adoption of rack and pinion steering, which is both sensitive and precise in action. The angle of roll during fast cornering has also been greatly reduced by the front and rear torsion bars. At fast touring speeds, the Cobra is easy to handle and the sheer luxury of having all that power in reserve is one of motoring's greatest sensual pleasures.

Driven to the limit, the Cobra is not for the beginner. To start with, it is a foot shorter in the wheelbase than other cars of comparable potency, such as the Ferrari, so one tends to be rather busy when drifting fast curves on full throttle. I admit that my test was too short to allow me to get to know the car, and I am sure that I would have gained confidence with experience. The machine is being improved with the rapidity which results from racing a new model, and I expect that it will become easier to handle in spite of its spectacular performance. In any case, it is only when driven at racing speeds that the Cobra demands considerable skill.

The traction afforded by the independent rear suspension, plus the self-locking differential, renders extremely rapid starts particularly easy to carry out. The disc brakes on all four wheels are immune from fading, but demand quite a lot of pedal pressure for emergency stops at fairly low speeds. Continuous use of the brakes from high speeds causes them no distress whatsoever, which is a comfort when so much performance is available.

Well-made and attractively finished, the aluminium body is of that excellence for which A.C.s are renowned. The hood does not flap, though it naturally bulges a little at over 120 m.p.h. American instruments were fitted to the test car with proper round faces, but the speedometer was inoperative so I calibrated the rev. counter, which was notably steady. I found the seating position very comfortable and the ride was good over all but the worst bumps.

The A.C. Cobra is a high-quality sports car with a stupendous performance. It is outstandingly flexible and has perfect town manners, while it may be driven everywhere on top gear or run at high speeds on the close indirect ratios, according to the whim of the moment. It is by no means heavy on petrol and its great reserve of power, when sensibly used, must be regarded as an excellent safety feature on crowded roads.

SPECIFICATION AND PERFORMANCE DATA

Car Tested: A.C. Cobra sports two-seater, price (America only) $5,995.

Engine: Eight cylinders 101.6 mm. x 72.9 mm. (4,727 c.c.). Pushrod-operated overhead valves. Compression ratio 9.2 to 1. 275 b.h.p. at 5,800 r.p.m. Four-choke downdraught carburetter. Two Ford ignition coils and Spalding distributor.

Transmission: Long single dry plate clutch. Borg-Warner four-speed all-synchromesh gearbox with short central lever, ratios 3.77, 5.29, 6.71 and 8.89 to 1. Hypoid final drive with self-locking differential.

Chassis: Ladder-type tubular frame. Independent front and rear suspension by transverse leaf springs and wishbones, with telescopic dampers and anti-roll torsion bars front and rear. Rack and pinion steering. Knock-on wire wheels fitted 6.50/6.70 x 15 ins. tyres. Disc brakes all round.

Equipment: 12-volt lighting and starting. Speedometer, rev.-counter, oil and water thermometers, oil pressure and fuel gauges, ammeter, clock. Flashing direction indicators. Heating and demisting. Windscreen wipers and washers.

Dimensions: Wheelbase 7 ft. 6 ins. Track (front) 4 ft. 3½ ins., (rear) 4 ft. 4½ ins. Overall length 12 ft. 7½ ins. Width 5 ft. 1 in. Turning circle 34 ft. Weight 18 cwt.

Performance: Maximum speed 136 m.p.h. Speeds in gears: 3rd 96 m.p.h., 2nd 75 m.p.h., 1st 56 m.p.h. Standing quarter-mile (damp road) 13.8 secs. Acceleration: 0-30 m.p.h. 2.5 secs., 0-50 m.p.h. 3.8 secs., 0-60 m.p.h. 5.2 secs., 0-80 m.p.h. 8.2 secs., 0-100 m.p.h. 13 secs.

Fuel Consumption: 14 to 18 m.p.g. (approx.).

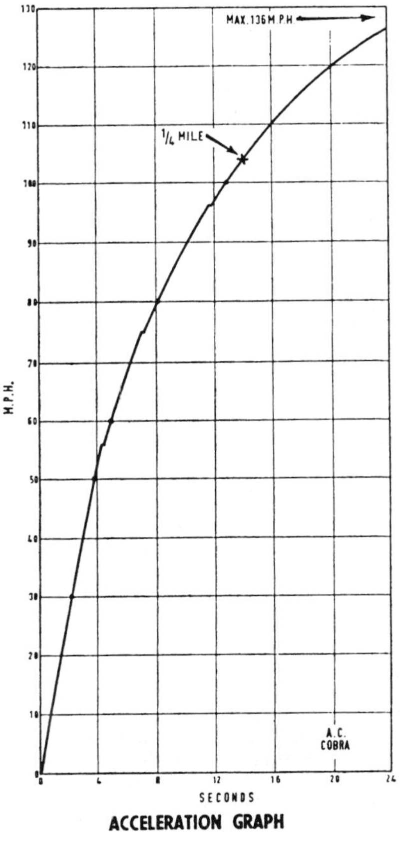

ACCELERATION GRAPH

Straight from Le Mans to M.I.R.A.; the Cobra still had its competition numbers. Light on the detachable hardtop was for identification after dark. The other team car had two.

UNCOILING THE COBRA

by Eric Dymock

CARE in handling Cobras may not be new to Reptile House keepers at Regent's Park. But to *The Motor's* road testers, Cobras are something exceptional; particularly in this case—a tuned, racing one; and hooded at that. The car which took seventh place at Le Mans was brought back from France, had its axle ratio changed, and was taken to the M.I.R.A. test track near Nuneaton. And what an exciting day's motoring it provided.

The Cobra is basically an A.C. Ace chassis with a 4.7-litre American Ford V-8 engine and Borg Warner gearbox. The brainchild of Carroll Shelby, who specialized in racing big, exciting sports cars in Europe and the United States in the late nineteen-fifties (he won Le Mans with Roy Salvadori in a DBR1/300 Aston Martin in 1959) the Cobra represents the highest development of the John Tojeiro-designed A.C. all-independent tubular frame. It is suitably strengthened for the Ford 289 engine producing, with the four double-choke 48/DMI Webers of the Le Mans car, around 330 b.h.p. With light-alloy body and detachable hardtop this car was driven at Le Mans by Ninian Sanderson and Peter Bolton, one of the handful of survivors and winner of the over 4,000 c.c. class at 108 m.p.h.

At first glance the Cobra is dominated by its wheels. The enormous 6.00—15 front and 7.00—15 rear Dunlop R.6 racing tyres seem to take up all the available wheel arch and bulge out the sides as well. They are mounted on alloy wheels to accommodate larger-than-standard discs; we drove the car with the Le Mans pressures of 50 lb. all round.

A.C. Cars Ltd., anticipating the possibility of using the car on circuits in this country changed the axle ratio from the Le Mans 3.07:1 to 3.77:1 before bringing it to M.I.R.A., but otherwise it was in exactly the condition in which it had crossed the line at 4 p.m. just eleven days before. Representatives of A.C. who accompanied the car said that the oil hadn't even been changed and the front tyres were those with which it started the race. They looked about three parts worn, but the back tyres were very nearly "gone." They were much nearer the "gone" by the end of the afternoon.

The inside of the Cobra is much smarter than most competition G.T. cars. Carpets were an astonishing luxury and the seats comfortable, but surprisingly the adjustment was inadequate for medium-height drivers to sit at arm's length from the wheel. This is perhaps less practical in the Cobra, however, because the steering is relatively heavy and could make straight arms tired. Constant reminders of the "racer" character, however, were the huge funnel in the roof behind the driver's head to the 30-gallon fuel tank and the shattering noise with the car at full cry. The facia is rather a clutter of small dials and one feels that three-quarters of Mulsanne must have been spent checking oil and water temperature, battery charge, oil pressure, fuel level, engine speed, etc. The standard speedometer worked but was obscured by the large spokes of the splendid, thick-rimmed, A.C.-manufactured, wood-rimmed steering wheel and it had not been compensated for changes in axle ratio which made it read 150 for 120 m.p.h. Not that the performance of this car needed flattering.

Using 6,500 r.p.m. on the very clear arc-type Rotunda tachometer, 40 m.p.h. could be reached in about 3 seconds; 60 in under 5; 80 in 8.1, and 100 in just 12 seconds. A further 6 seconds or so took us past 120 before the straight gave out and the time for the standing ¼-mile was 11 seconds with a terminal speed of almost 107 m.p.h. *still in third gear*. On the banked high speed circuit, the Cobra went through the electronic timing strip at 139.6 m.p.h. at about 6,200 r.p.m. but would gain several hundred more revs before it was necessary to slow down for the next banking and one felt that it would easily pull over 6,500 on a slightly longer straight,

representing about 148 m.p.h. With the Le Mans axle, it was doing about 160 m.p.h. at 5,500 r.p.m. on Mulsanne.

Performance of this sort is electrifying, but the handling of the car made it unexpectedly easy to cope with. The figures disappointed the people from A.C., probably owing to a flat spot around 3,500 r.p.m. because of the need for some engine re-adjustment. But it should be remembered that nothing had been touched since the completion of 2,592 very hard racing miles at a speed high enough to have won the 1956 (sports car) race when, incidentally, Ninian Sanderson, co-driving with the late Ron Flockhart did in fact win with an Ecurie Ecosse D-Type Jaguar.

On the road circuit at M.I.R.A., the Cobra exhibited powerful, consistent understeer but the adhesion was so good that on a long bend it was only necessary to turn on a little lock and a lot of the immense power to set it up in a stable, predictable drift. Even tight corners in the lower gears could be taken with a good deal of power on; in fact it was necessary to keep the throttle open to keep the front on the line one wanted. There was quite a lot of roll for a competition car despite the heavy anti-roll bars

Power Station. One carburetter choke per cylinder consumed fuel at Le Mans at around 8 m.p.g. A spare fan belt is still taped up beside the radiator header tank.

but the frame felt stiff and taut, even at high speed on the rather bumpy banking.

The Borg Warner gearbox turned out to be unexpectedly good, with a rapid change and powerful synchromesh but the ratios are rather wide. The short lever needs a firm movement together with a clutch which did not appear unduly heavy but took unkindly to a succession of standing starts. Apart from a strong smell of hot lining, however, it did not appear to suffer any permanent disablement. It was a difficult car to get off the line on account of the flat spot, and one was faced with the choice of letting the clutch in at around 6,000 r.p.m. and coping with tremendous wheelspin, or starting at a lower engine speed and having the revs die momentarily below the all-important 3,500. The large disc brakes, with racing pads and heavy pedal pressures steadfastly refused to show the slightest signs of fade, even under the stress of stopping the car from over 120 m.p.h. every two or three minutes while we took the acceleration figures.

The Le Mans Cobra is large, powerful, exciting. It has the high-scuttled, massive feel of an earlier era with the stiff-framed precision and comfortable ride of a modern thoroughbred. Driving Derek Hurlock's own road Cobra the same day showed that the deep, braying, exhaust note of the big V-8 could be allied to great flexibility and acceleration which is unobtrusive but immensely swift; not kick-in-the-back style, but more like a rocket leaving the launching pad, and going on, and on, and on . . .

Production of Cobras for the British market is held back only by the supply of engines from Ford of America. When these become available, this car is bound to repeat its Le Mans success —and succeed.

Executive's office of a Le Mans class-winner. The Rotunda arc-type tachometer is mounted on the passenger's side and the big steering wheel has a comfortable, thick rim. Foam pad on right is for the driver to rest his knee against.

Start of the A, B and C class production race and the Cobras of Johnson, MacDonald and Miles are out front (where they finished).

THE COBRA STRIKES AGAIN

STORY BY FRANK H. BENNETT

PHOTOS BY HERB WILLIAMS

IN THE ARTICLE, "Corvette vs. Cobra" (R&T, June 1963), the author stated: ". . . it is very hard to imagine that any well prepared, well driven Cobra will be beaten this year —not by the Corvettes, and possibly not by anyone, unless the organizers get sneaky and push the Cobras over into the same races with all-out racing cars. There are, as a matter of fact, rumors of this happening, and if it does, the Cobras just might beat the big modified cars, too."

Well, it happened! It happened at Lake Garnett in the National SCCA racing event of July 7.

It happened in a grand and spectacular manner, and 60,000 spectators saw the biggest racing show of their lives!

The Shelby-Cobras came to Garnett, Kan., three of them. With them came Carroll Shelby himself, the man who developed this racing machine. The drivers he brought were Bob Johnson of Columbus, Ohio, Dave MacDonald, El Monte, Calif., and Ken Miles of Hollywood, Calif.

Also at Lake Garnett was the famed Grady Davis Corvette team, which includes Dick Thompson, Washington, D.C., Don Yenko, Canonsburg, Pa., and Davis himself from Pittsburgh. They brought four cars: a couple of A-Production Corvette Sting Rays, a B-Production Corvette and,

Start of the modified event; Hinkle's Cooper Monaco is in front. Miles (98) won, followed by Hinkle and Heuer's Chaparral (12).

A smile from Miles—one driver who really enjoys himself.

"Here lies the Sting Rays" reads the grave marker; a whimsical touch from the Cobra crews. Could it be that the needle is in?

Ol' Shel, barely recognizable without his coveralls, wears the "bad guys" black hat. Miles is shirtless, a mechanic ponders.

rounding out the card, a C-Modified Corvette Grand Sport.

And there it was, the showdown everyone had anticipated, with the answer to everyone's question—What would happen in a race between Cobras and Corvettes? And how would both of these American race cars fare on the challenging 2.8-mile Lake Garnett road racing course, that curls like a sea-horse around Garnett's 65-acre lake?

A few minutes after 1 P.M., the A, B and C Production cars were given the green flag. In less than half a mile, the Cobras had taken a substantial lead over the entire field, a lead which they continually increased during the 45-minute race. They finished first, second and third—Bob Johnson, Dave MacDonald, Ken Miles. The Corvettes came in fourth, fifth and sixth—Dick Thompson, Grady Davis, Don Yenko.

Before running in the production race, Ken Miles' Cobra had been challenged as being modified by an oil cooler. Mechanics hurriedly removed the cooler for the production race. Then Miles and Shelby reasoned that if the removal of the cooler put the Cobra in the production class, then the return of the cooler to the car would put it in the modified class. Officials agreed, and Miles entered the feature race: C, D, E, F and G Modified. And at Lake Garnett a Cobra finally showed its pipes to the big modifieds!

Miles ran away from the field. He drove a superb race, lapping the tricky Lake Garnett course 23 times in 45 minutes and averaging more than 85 mph, to win the Lake Garnett Grand Prix overall championship for 1963.

Harry Heuer in a Chaparral came in second and Jack Hinkle was third, in a Cooper Monaco. Dick Thompson, driving a Corvette Grand Sport, was among the dnf's, and while he was in the race he trailed the Cobra all the way. At one time, he was in second place, but as for his chances of outrunning the Cobra, 60,000 race fans will tell you that it simply couldn't have been done at Lake Garnett.

Map of the circuit. A car went into the lake during main event.

SHELBY COBRA --POWERED BY FORD

An ANGLO-AMERICAN CHALLENGE

Carroll Shelby has taken three major automotive elements, an ocean and half a continent apart, and compressed them into one of the most potent sports cars on today's racing scene. Using the well designed body and chassis from AC of England, and the husky 289 cubic inch engine from Ford, he has developed an honest-to-goodness grand turismo car that is capable of cruising well above the 100 mph mark with ultimate ease and security.

How the Cobra can be so servile in its lower ranges, and so all out hairy on the high end of the clock is better experienced than talked about, but it combines the handling characteristics of a good four wheel independently suspended chassis with a torque-laden V8 engine in a way that is guaranteed to please the most finicky high-performance buff.

The chassis follows the design of the standard AC "Ace," with two large diameter tubes serving as the main structural members. Between these, on each end, are box section units that serve as anchoring points for the suspension links and springs. Also tied into the main chassis members is a network of tubing that becomes the support for the aluminum body shell.

Shelby has lengthened the A arms and the transverse leaf springs, thus widening the track, and necessitating the use of distinctive flaring around the wheel wells. The suspension has also been stiffened by the addition of an extra spring above the main leaf, and to reduce oversteering, negative camber has been introduced in the rear wheels.

Our driving impressions of the Cobra were obtained under ideal conditions, at the Riverside Raceway, where it was possible to see firsthand how well the car responds to the driver's wishes. Even with the standard engine, the acceleration is startling, and the wonderfully spaced gear ratios made the whole operation no more difficult than driving the family car around the block. Yet 100 mph comes up with alarming quickness, and the Girling disc brakes, 12 inches in diameter, have speed reduction characteristics that more than match the engine's acceleration capacities. We took a ride in one of the team cars, with Ol' Shel himself at the wheel, and among the things that impressed us most was the fantastic way that the brakes hauled us down from 150 mph plus prior to entering turn 9. There was not the slightest trace of uneveness as the Cobra slowed to a docile

90 mph, then he brought the car to a complete stop in the pit area. The sensation when he first applied the brakes must be very similar to that of what an astronaut feels when he fires the retro fire sequence in his capsule. A momentary negative G load builds up, and finally equalizes, as the speed of the vehicle is reduced.

I must say that our own driving efforts were psychologically hampered by having the likes of Ken Miles sitting in the passenger seat. Not that he placed any restrictions on what we did, but after all, one does not like making a fool of oneself in the presence of such talent. His restraint was admirable, and British upper lips were never stiffer.

Down in the engine room, there is little unused space. The lightweight Fairlane V8 is mounted well aft of the front wheels, resulting in a weight distribution of 48/52%. In its stock or street guise, the engine is fitted with a single four barrel carburetor and produces 270 hp at 5800 rpm. It is virtually a stock Fairlane engine, except that it has solid valve lifters, and a reground camshaft.

When set up for racing, the engine uses four double-throat Weber carburetors, a camshaft ground to give maximum top end performance, and a higher compression ratio. In this state of tune. something in the order of 325 hp is available, and if you think the stock version is something, you should get a load of this model! It was in a Cobra of this configuration that we took the hot lap, with Shelby driving. It made believers out of us.

Perhaps the most significant thing in recent Cobra history is its fine showing in the Le Mans 24 Hour Race. Granted there were six Ferraris ahead of it at the finish, but to finish 7th in a field like that is something to be proud of. It is the first time in a very long while that an American production engine has been so well placed.

All of the drive line elements are heavier than those used in the standard AC. The transmission is made by Borg-Warner, and is four speed. In the earlier models the

289 cubic inch Ford Fairlane engine fits neatly under the hood.

Neat instrument cluster in the roomy cockpit.

The Cobra at Le Mans. It finished 7th behind six Ferraris.

The husky Girling disc brake, as mounted on the Cobras.

The chassis and body framework features strength and light weight.

SHELBY COBRA

"gap" between third and fourth gear was somewhat wide, but it has now been reduced by new tooling. Because of the greater power available from the Ford engine, all of the final drive members are heavier than those in the AC. The Cobra utilizes Salisbury drive-gear assembly to get power to the rear wheels through Spicer shafts and U-joints.

The instant one sits in the driver's seat he knows he's in a race car. Bucket seats, leather covered, give positive support, and the seating position and the proximity of the gear change lever are close to perfect. As complete range of instruments, including a tachometer, speedometer, oil pressure and temperature, water temperature, clock, ammeter and fuel gauge, are nicely clustered within easy view. Standard equipment includes full weather protection, with sliding side windows, and full tonneau cover.

This entire car reflects the careful planning of a man who is really aware of what is needed in a competition automobile, and at the same time knows the necessity of certain animal comforts. The finish of the Cobra is equal to or better than any car of its type that we have had the opportunity to examine, and we do not hesitate to predict that the racing future of this car is bright indeed.

Standard four barrel manifold.

Four Weber carburetors as used on the competition cars.

MS

ROAD TEST

COBRA

WEIGHTS AND MEASURES

Advertised weight	2030
Weight as tested	2412
% on front wheels	49%
% on back wheels	51%
Length overall	151.5 in.
Height	49.0 in.
Width	61.0 in.
Ground clearance	7.0 in.

BRAKES

Type	Girling disc
Area (square inches)	580

PERFORMANCE

Top Speed (manufacturers figure)	155 mph
Average of several runs	
0-40	3.0 sec.
0-50	3.8 sec.
0-60	4.2 sec.
0-70	5.6 sec.
0-80	6.9 sec.
0-100	10.6 sec.
¼ mile acceleration from standing start	13.8 sec.
Maximum speed at ¼ mile	112
Gasoline consumption—overall average	12.6 mpg

ENGINE

Type	V-8
Bore	4.0 in.
Stroke	2.87 in.
Displacement	289
Compression ratio	11 to 1
Torque	not available
Horsepower	270 at 5800 RPM

REAR AXLE RATIO	3.77-1

ILLUSTRATIONS BY TOM QUINN

Fast driving is transformed from dark mystique into cold logic during five long days at Riverside

The classroom at turn nine

BY JOHN JEROME

There are approximately a quarter of a million people who read *Car and Driver* every month—a quarter of a million people who are interested in driving, who love cars, are conversant with the techniques involved in racing, and understand the basic physics of high speed in automobiles. There are maybe twenty out of that group who wouldn't gain by taking the driving course offered by Carroll Shelby's School of High Performance Driving.

If that sounds like a deliberate insult, it is. Of course I realize that *you* are one of the twenty—just as 98% of the nation's drivers characterize their own level of driving skill as "expert," and estimate the level of all the rest of us as bloody awful. At least they are correct on the last count—for that matter, you may be correct in including yourself among the hypothetical twenty readers that I just dreamed up who wouldn't derive substantial benefit from taking the course.

I personally went to take the course as a reporter, not a student. Hell, why should I consider myself a student when I had 10 years of off-and-on high speed driving experience behind me, an average of 50,000 highway miles a year for that period, a racing background that was reasonably successful (if spotty and sporadic)? I had to admit I was a super-sophisticated magazine writer, and I had one hell of a lot of faith in my ability as a driver. I got over all that quickly enough—not in any embarrassing fiasco (that came later), but in a succession of driving sessions in which I was straining—but not ever quite able—to cope. I learned an awful lot about driving, it seems to me now, and I shudder to think of some of the things I had previously attempted. And I came away completely sold on the school, its instructors, their method, the good they accomplish, and the efficacy with which they impart their background.

What is so great about the school is that it provides the novice or semi-novice with a means of reducing the problem to its smallest elements, learning those elements thoroughly, and then adding back at his own pace all the contradictory elements, distractions, considerations, and confusions of the racing or high-speed driving situation—at a rate, and in an order, so that coping with them is a natural progression.

For example: the first driving exercise involves Turn Nine at Riverside, a constant-radius, slightly-banked, medium-speed turn with a very solid vertical wall on the outside. The method of approach is to go slowly through the turn as a passenger with an instructor driving, who gradually picks up speed until he is going through fairly well hung-out. Then the student is turned loose on the turn. A recommended line is pointed out, and a recommended apex marked.

You may well spend an entire *day* on that turn. Everyone has a fairly consistent idea of what the correct line is; nearly everyone has a little trouble getting the car to follow that line. But the turn itself is a "constant" enough problem that you need worry, at first, about only two controls—steering wheel and throttle. You aren't going fast enough to worry about braking, and you just stay in third gear (in the school's street Cobra).

You begin by approaching the turn cautiously, of course. The possibilities in regard to line seem very limited, and the wall seems very close. Eventually, after a couple of dozen trips through the turn (it's one end of a quarter-mile oval, another reason for beginning there—you circulate around the oval and give Turn Nine another crack before you forget what you did last time, ideally), the terrain begins to work itself out in your mind. You begin to pick up individual differences in patches of pavement, particular markers that will assist you in picking your way through. The pavement begins to widen for you, and the possibilities for proper line widen at the same time. As a matter of fact, the ways to foul up the corner multiply for you at almost the same rate that your skill develops. Finally you feel you have it knocked, and you are able to enter the turn where you want to, come out where you want to, clip the apex wherever you choose, maintain whatever attitude seems best at the time—and you are *still* a second slower than the instructors.

You know this because of one of the basic principles of the school. Every attempt at any turn, by student or instructor, is timed. On each turn, an entry point and an exit point are designated, and the car is timed each time it goes through. Now, 30 laps after you begin, the instructors may call you in and have you watch from the apex while one of them goes through a half-dozen laps. Or one of them may take you through as a passenger. Both treatments help—as your concept of the corner broadens, your ability to spectate, to interpret what is going on out on the course with the car, broadens also (even for a guy who's been reporting races for 10 years). And you are getting a continual set of comparative lap times.

To back up a bit, the actual driving only begins after the student has enough "ground school" so that the instructors have a fairly clear notion of how much he knows about vehicle dynamics. A particular point of view—in effect, the philosophy of the entire school session—crops up at this point, and it takes a little time and a lot of discussion before the neophyte gets familiar with thinking in the school's terms. Pete Brock, the able designer who is most responsible for working out the curriculum of the school and who is chief instructor, likes to talk about racing almost entirely in terms of the contact patches where the tires touch the road. Since whatever finally occurs with the car is directly dependent on what occurs with those patches, there is hardly a better place to begin thinking about what is going on, how to improve performance, how to achieve the desired effect. That is the one place where stability is imparted to the car, and the one place where, if it is lost, control is lost.

In these terms, everything the driver does must stand in terms of that contact with the road. Whatever is going on with clutch, brake, steering wheel, shift lever, and intestinal fortitude is only a means to the end of ringing variations on what happens with those four patches of contact with the road. If it happens that

you come to the school already capable of doing pretty much anything you want with the aforementioned controls of the car, great. You are a little bit ahead, anyway, and can further simplify the problem because you aren't trying to learn to *drive* while you are trying to learn to drive *fast*.

Once you understand the importance of that contact, you can begin to think in terms of what effect the motions you make as a driver have on those contact patches. You are immediately swept up in a swirl of effect and countereffect, the subtleties of smoothly transferring weight, say, from both front wheels to outside front, then to outside front and rear, then, by application of power, to outside rear—and what all that does to the tires' capability for handling side loadings and slip angles. You get involved in the complexities of what a stab on the brakes can do—for you or against you—or, more subtly, what turning the wheels can do: the weight transfer forward as well as sideways, and the effect that that is apt to have.

A case could be made that this is more complicated than it is. Basically the principles are simple (loading a tire increases its slip angle until the maximum is exceeded, then the tire breaks loose—for example), and once you get used to discussing car behavior on a given turn the ideas involved are a lot clearer. Then you graduate upward into sophistication—upward, literally, from the basic simplicity of the patch of rubber on the asphalt, through the complication of the pneumatic carcass and the wheel, then into the suspension itself and what loading, preloading, and unloading do to vehicle dynamics. It *could* sound complicated just talking about it.

But getting back to Turn Nine—I was just getting confident about it all when I discovered—thanks to a brief demonstration from Pete Brock—that for all that sublime precision control I was suddenly able to impart to the matter, I was still a second slow. Then I began hurrying, and began learning some neat lessons about racing For one thing, I picked up my entry speed for Turn Nine by nearly ten mph, and my exit speed by a good five mph. I was no faster. I was a lot more sideways, and a lot more nervous about the whole thing, but I wasn't getting close. I ended up, for example, spending a whole day on Turn Nine, raising my exit rpm by about 500. Three days later, after I had worked on other turns and was pretty well familiar with the whole course, I found that I'd picked up another 500 rpm in exit speed without consciously working at it—the things I was learning were beginning to sink in.

The procedure, then, is a basic schooling in vehicle dynamics, long and hard practice on a very simple turn until the blackboard sessions begin to sort themselves out in the students' head, and then on to more complex problems. Every ten laps or so there is some discussion. *Why* you aren't getting around as fast as you should be is pointed out—it may be line, it may be improperly accomplished weight transfer, it may be that you simply aren't driving fast enough. You are encouraged to experiment, particularly after the first day or so, when you are comfortable in the car and the instructors are reasonably comfortable about your basic skill. You are also encouraged to hang it out. Brock wants you to learn to get the car committed, to push the car's suspension and tires to their limits so that you know what the limit feels like, what is irrevocable and what is merely over-exuberant. They want you to be able to hang it out, safely and in a consistent fashion, so you then can begin working back toward maintaining bite and getting the power to the ground, not sliding the tires so much that you scrub off speed and rpm. They don't want you to be hairy. But they do want you to find out what a car can do, and they'll push you in that direction.

So the complexity is continually increased. You may get a bit disgusted with Turn Nine before you are through with it, but by the time you *are* through with it you are ready to progress at a rate that will effectively prevent your getting bored with any of the rest of the course or the school. For the rest of Riverside, the other eight turns have distracting elements like braking and shifting thrown in.

One of the most remarkable developments, as far as I was concerned, was my memory—not for the course or the turn, but for what I'd done previously. The first

50 times I went through Turn Nine, for example, were pretty much a mindless blur, and I'd be trying so hard to remember the previous attempt that I'd foul up the next one, through lack of concentration. Finally, with the accumulation of time, I began to be able to remember. Particularly with the aid of the constant time-taking, I finally found myself able to run ten laps and then come in and check times, able to then link successive lap times with successive laps and remember how I'd fouled up. I began to get a first glimmer of how a good racing driver could become able to lap consistently on a road course to within a few tenths of a second. And that in fact is the stated purpose of the school: not to make Walter Hansgens of us all, but to get the students so that they are capable of lapping a road course at a comfortable pace, lap after lap, without varying seconds either way—and without the risk of doing something completely stupid.

Pete Brock is a bright, articulate young Southern Californian with too many talents for the hours in a day. He designed the Shelby Cobra coupe which led the Daytona Continental until fire in the pits knocked it out, recently won the Grand Touring category at Sebring, and set a GT lap record at Spa. He is in charge of special projects for Shelby, which seems to include everything from designing letterheads to teaching the school, designing show cars to art-directing advertising campaigns. He's effective enough as a racing driver—he's one of the better small-modified drivers in Southern California, and gets an occasional stock car road race ride. Carroll Shelby frequently enters a Cobra for him in the larger races—then Dan Gurney usually comes along and Pete gets bumped at the last minute.

John Timanus is "Mr. Lotus" on the West Coast, probably the most-experienced small-modified driver in the area. He's a big, easy-going fellow with a completely calm manner of hurrying cars about the course. Not so articulate as Pete, he's also more inclined to experiment. Between the two of them, Pete and John have so much experience at Riverside that a six-inch variation in the established line on any of the turns is immediately apparent—and significant—to them.

Attending the school is faintly reminiscent of an athletic training camp. The pace is leisurely but monastic, with four or five hours a day of driving and a lot of time to sit poolside at the motel thinking about it all. You find the sessions impinging on your consciousness continually while you are away from the course, and you tend to daydream about what you were doing wrong and how you'll improve the next day. If your car is of large displacement, and you aren't in shape, you'll be sore for the first couple of days, and probably terribly proud of the Phil Hill-like blisters on the palms of your hands.

It would seem to me, now, that a beginner wanting to get into racing ought to plan on three things. He ought to spend a couple of years hanging around learning about preparing racing cars—there's no better way than just picking that up by being there. He should attend a couple of SCCA drivers' schools, because the kind of information about safety and traffic and racing etiquette (yes, seriously) gets across better there than anywhere else. Then he ought to save his pennies and his time, and haul his car to the Shelby school.

The cost is $500 with your own car (which you ought to have for the course—and plan on spending some more money on tires, because you run a lot of fast miles), or $1000 in a school car. If you have the usual experience and no car, you use a school Sprite; if you are an accomplished driver, you may get a Cobra. Yes, it is expensive. When you see the treatment, the personal instruction, the man-hours and equipment-hours involved, you'll be more than satisfied that you are getting your money's worth—in fact you'll wonder how they do it for that price. Besides, you'll *make* it worth the money and the time. It is also a good idea to plan to go alone, with no other purpose for the trip.

The standard wisecrack about graduates of the school runs along the lines: "Well, I don't know if it's the instruction or five days of running around Riverside, but ol' so-and-so sure is going a lot better than he did last year." The crack gets made because ol' so-and-so is always going much better. But the practice-versus-instruction part of it is inseparable, as it should be. You can't get that kind of intensive, analytical practice on any race course anywhere. And to get it in concert with continual running criticism from knowledgeable observers, with each component of each lap timed, and with a basic background carefully laid to prevent the possibility of going at it all wrong, is almost more than a driver can ask.

More than that, though, the student is getting a demonstration of the application of the dynamics that affect high-speed vehicles, and in a more vivid and dramatic fashion than he could get any other way—with himself at the controls. He's in effect the subject of a laboratory experiment—under as carefully-controlled conditions as possible, an experiment in seeing how much a guy can learn about making a car go fast. C/D

AC COBRA

WORLD'S FASTEST PRODUCTION CAR?

By MIKE McCARTHY

Tall Texan Carroll Shelby's AC-Cobra is a real firecracker destined to go places fast

Photograph: Pete Brock

Even though Cobra's performance is second to none, car is still extremely practical.

Carroll Shelby hopes to roll 350 AC Cobras off his small production line this year.

WHILE it has not laid official claim to the title, recently conceived Anglo-American AC Cobra is surely the world's fastest-accelerating production sports car and could undoubtedly give many out-and-out competition cars a run for their money. This dual purpose machine in stock trim accelerates from 0 to 100 in just 10.8 seconds, which is really motoring by today's standards. Yet, the AC Cobra hides nothing particularly exciting under its attractive bodywork — power to weight seems the secret of its devastating performance.

The man responsible for the birth of this remarkable machine is the tall Texan racing driver, Carroll Shelby, whose concept called for the type of car that has all but vanished. It was to have sheer road performance second to none, yet it had to possess those elusive characteristics which would make it equally desirable for both normal road work and racing. Above all it had to be thoroughly practical.

The first thoughts about this remarkable car were born when Shelby was at the peak of his racing career. In 1956 and 1957 he was the Sports Car Club of America champion. He followed this up by sharing the winning car in the 1959 Le Mans 24 hour race and winning the United States Automobile Club road racing championship in 1960.

Working in collaboration with the Ford Motor Company and AC, Shelby had a prototype built to his specifications early in 1962. At that time the intended engine was the 221 cubic inch Ford V8 from the small Fairlane. The basic frame, suspension and bodywork came from the long-established English firm, AC. It was derived from the AC Ace two-seater designed by John Tojiero in the early 1950's and since then has continued more or less unchanged.

The ladder-type frame comprises two large diameter tubes while a simple independent suspension of upper transverse leaf spring and lower wishbones is employed on all four wheels. Driving axles are of the sliding splined type.

To suit Shelby's needs, AC strengthened the frame and various suspension components. Stiffer springs and dampers were fitted, as was a stronger final drive assembly. The differential is the same Salisbury unit used in the E-Type Jaguar. The track was widened by four inches and 15 in diameter 72-spoke Dunlop wire wheels replaced the original 16 in wheels. Discs are fitted all round, being 11.5 in diameter on the front and 10.5 in on the rear. There are single calipers on the front and double calipers on the rear.

External changes were minimal. The aluminium body was flared at the wheel arches to accommodate the increased track and the nose restyled for improved penetration and appearanee. Drag, and buffeting in the cockpit, were lessened by fitting a windscreen with more rake and curve than that on the Ace.

The car, in a semi-finished state, was shipped to Shelby's firm in Califomia, for installation of the engine and transmission. By the time all the revision was finished, however, the Ford Motor Co had announced an optional 260 cubic inch version of the lightweight 221. And it had also developed another engine producing 260 bhp. Ford took the basic 260 engine and gave it 10 to 1 compression ratio, larger ports and valves, a mechanical valve train, dual point ignition and a four-throat carburettor.

AC COBRA WORLD'S FASTEST PRODUCTION CAR?

. . . *Continued*

Photograph: Pete Brock

Above: This, the world's fastest-accelerating production sports car, possesses Jekyll and Hyde characteristics. It can be driven to work and win races on weekends.

Left: One of Shelby's four team Cobras caught in the process of setting a new class lap record at America's Riverside Raceway.

Photograph: Friedman

Shelby slipped the new engine into place and had himself a rocket! The only difference between the Cobra's engine and the stock Ford unit is that the Cobra version has fabricated tube exhaust headers. Behind the engine is a four speed, all synchromesh Borg Warner gearbox of the same type used in the Chevrolet Corvette. The ratios however, are slightly different.

The car was a resounding success right from the time it was first fired up. It wasn't long before Shelby was able to confidently advertise "Buy it, or watch it go by". This is no idle claim. The Cobra is unquestionably the fastest accelerating sports car in the world today — and it has road behavior of a high order.

Yet, despite its performance, nothing is lost from the car's habitability.

The cockpit is as thoroughly functional as it is simple. From the driveshaft tunnel protrudes a stubby gear lever. Ahead of the driver is an adjustable steering column and an alloy spoked, wood-trimmed wheel. Immediately in front of this, on the flat leather-faced dash panel, is as comprehensive a set of instruments as anyone could wish for — a 180 mph speedometer, an 800 rpm tachometer — redlined at 7200 rpm — and gauges for oil pressure, amps, oil and water temperature. All-weather protection is provided by a folding soft top and side curtains with sliding windows. The boot, even though housing an 18 gallon fuel tank, is particularly commodious.

In **standard** trim the Cobra is in a relatively mild state of tune, considering the long list of options. Most of these relate to the transmission and final drive ratios, and to a selection of hot-up equipment for the engine. There are four distinctly different carburettors. A single four-throat downdraft carburettor is standard equipment, but this may be supplanted by two four-throaters, three twin-chokers, or (for purely competition purposes) four twin-choke Webers on a ram manifold.

Shelby also hopes to make GT cars available in the US based on AC's Aceca. Coupled with the hot version of the Ford V8, producing more than 325 bhp, the car will probably be even faster than the Cobra because of the streamlining advantage.

By ultilising the most efficient induction system, and specifying a hotter cam and higher compression, a Cobra owner can avail himself of up to 330 bhp from the 260 cubic inch engine. Should he be courageous enough to want even more power and torque he can have a full-house 300 cubic inch Ford engine. When worked over this unit can produce 350 bhp!

Problem: If a 260 bhp Cobra does 0-100 in 10.8 sec what would be its time with 350 bhp under the bonnet?

COMPARISONS IN PERFORMANCE

	COBRA	CORVETTE	E TYPE
Engine	V8, pushrod ohv	V8, pushrod ohv	In-line six, dohc
Bore	3.80 inches	4.0 inches	3.34 inches
Stroke	2.87 inches	3.25 inches	4.17 inches
Capacity	4262 cc.	5359 cc.	3781 cc
Induction	one four-throat carb	fuel injection	three SU carburettors
BHP	260 @ 6500 rpm	360 @ 6000 rpm	265 @ 5500
Torque	269 pounds/feet	352 pounds/feet	260 pounds/feet
GENERAL SPECIFICATIONS AND DIMENSIONS			
Wheelbase	90 inches	98 inches	96 inches
Weight	2020 pounds	3036 pounds	2720 pounds
Pounds per bhp	7.7	8.4	10.2
PERFORMANCE			
0.40 mph	2.5 sec	3.0	4.6
0-60 mph	4.1 sec	5.6	7.4
0-80 mph	6.8 sec	10.1	11.6
0-100 mph	10.8 sec	14.1	16.7
Standing ¼ mile	12.9 sec	14.1	15.2
Speed at finish of ¼-mile	114 mph	100 mph	94 mph

COBRA SPECIFICATIONS AND DIMENSIONS

Wheelbase	90 inches
Track	51.5 inches front, 52.5 inches rear
Overall length	151.5 inches
Width	61.0 inches
Height	49.0 inches
Frontal area	16.6 square feet
Ground clearance	7.0 inches
Turns, lock to lock	two
Turning circle	34 feet
Kerb weight	2020 pounds
Weight distribution	48/52
Tyre size	6.70 x 15
Mph per 1000 rpm (4th gear)	21.8 mph
Mph in gears	first (7200 rpm) 67 mph second (7200 rpm) 89 mph third (7200 rpm) 112 mph fourth (7000 rpm) 153 mph

Less sporting types will probably prefer the heater/demister and radio that are also options.

Judging from the praise lavished on the Cobra by American road testers, the hot hybrid's performance and handling characteristics apparently belittle the majority of superlative terms. We must assume that it approaches the ultimate. Yet the Cobra is as docile as any **boulevarde** sports car.

Although the production rate could have been much higher than the 350 envisaged for 1963, Shelby insists that he and his team build only as many cars as they can give their full attention. Every car is personally shaken-down and tested by Shelby as part of the rigorous final inspection. Probably the only people disappointed in the Cobra are those whose money-in-hand orders could not be accepted because the first year's production quota was sold almost before the doors of the 1962 New York Automobile Show (where it was introduced) swung open.

Apart from those tantalised persons, however, everyone is happy about the Cobra — Shelby, AC and Ford because of the car's success, and the enthusiasts because it represents a marked step forward in the evolution of the sports car.

In the 1963 12-hour race for sports and GT cars held at Sebring, Florida, several AC Cobras were entered all in various stages of tune. They were the only cars to challenge Ferrari during the first hours of the race, blasting down the straights far faster than anything else, but handling like elephants with greasy back feet on the corners. Apparently the fantastic power these cars were pushing out had an affect on transmissions and suspensions after a few hours racing.

Graham Hill and Dan Gurney were lying in fourth position at the four-hour mark until the steering finally gave up when the car was lying fifth during the sixth hour.

It is undoubtedly a car with a great future and once handling problems are sorted out the continentals will have to watch every move they make or it will be an AC Shelby Cobra all the way. #

ABOVE – Comparison of stock head, left, and Cobra component indicates how valve pockets have been relieved in Dragonsnake to permit freer breathing. Heads were also milled, upping compression from 9.5 to 10.5 to 1. All chambers measure out exactly to 42cc. Stock 1¾-in. exhaust and 1½-in. intake are now 1⅞-in., 1⅝.

ABOVE RIGHT – Handheld flat-top piston is production item that is exchanged in the hotted-up version for slugs with four valve pockets machined in the crown. This is necessitated by close proximity of valves after head mill.

Off and running, usually well ahead of the competition, driver Ed Terry storms across the opposite page. Blue metalflake machine was senastion at the Nationals, progressing downward from 12.32 e.t. at first to high 11's to win A/Sports class. All dressed up in its Sunday best, 289 mill indicates what can be accomplished with some tasteful chroming, paint and a set of those sought-after Weber carbs. Standard accessories include Autolite silicon plug wires and Wedge scattershield.

Quick Cobra

Shelby American "Dragonsnake" puts the bite on records and its rivals

photos by Eric Rickman

by Eric Dahlquist ■ In this sporting life of drag racing, there appears on stage occasionally a car that electrifies competitors and crowds alike, blazing a trail across everyone's memory with the acuteness of its performance. Sprinkled lightly throughout the profusion of machinery running in the nation, the uniqueness of this special breed lofts it above the multitude, chiseling a secure niche in the quarter-mile chronicle. The Cobra Dragonsnake is such an animal. Off the line in a nose-high attitude, Cure-Ride 50-50's locked in position, the Anglo-Americanized azure sportster issued its first challenge at the HRM Championship Drags, going on to clip both ends of the NHRA A/Stock Sports record with 116.27 mph in 11.81 seconds. Later, Ed Terry grasped the reins of the Shelby American sizzler, tracking class rivals to their lairs on a country-wide tour, and turned them back at every meeting. With the fraternity alerted, Terry towed to Indianapolis for the supreme trial under fire: the Nationals. Pitted against the cream of the crop, and before the critical eyes of thousands, he ripped away from every challenge in a crammed class, erasing all doubts as to who is the tiger of sports hill in drag racing. ■ ■

and aluminum body made in England by AC Cars.

At the time, Shelby was working out of a small office in Santa Fe Springs, Calif. In the spring of 1962, Shelby's prototype Cobra was ready. The name "Cobra" had come to Shelby in a dream. He awoke and wrote the name on a pad beside his bed, deciding the next morning this was the name he would use for his car.

The first AC bodies were delivered to Santa Fe Springs for production in March, 1962. Before many cars were completed, the factory was moved to Venice, Calif., where Shelby had leased the shops once used by Lance Reventlow to build Scarab sports cars.

World's Fastest Production Sports Car

The 260-inch Ford Fairlane engine originally used in the Cobra gave way to the 289-inch version, as the Cobra became the world's fastest production sports car.

In its first year of production and first season of competition, the Cobra won the United States Road Racing manufacturers' championship and Cobra drivers won the USRRC and SCCA Class A Production racing titles.

The 1963 season was dominated by the Cooper Ford Cobra, a development of Shelby's based on the same engine that powered his production car.

The 289-inch Ford-powered Cobra in 1964 won its second consecutive USRRC manufacturers' championship and finished within seven points of Ferrari for the world title. Shelby's new version, the Cobra 427, is now in production. It incorporates a new 427-inch light-weight Ford engine, coil spring suspension, and greater length and width than the previous Cobra.

The Shelby American company has recently moved into a 96,000 sq. ft. facility at the Los Angeles International Airport, where the Mustang GT 350 and future Cobras will be manufactured. ■

DOUBLE TEST: 427 COBRAS FOR STREET & COMPETITION

BY JOHN CHRISTY & JERRY TITUS

AS A NUMBER OF PEOPLE HAVE FOUND OUT, when you have a world beater on your hands, be it automobile or outhouse, you don't just sit back on your laurels and expect people to let you get away with it.

So it was, and is, with Carroll Shelby and the Cobra. In 1964 and 1965, Shelby had the Production car world by the tail with his 289 Ford-powered "snake." Except in very, very rare instances, where all-out effort was coupled with all-out driving ability, the Cobra was unbeatable. It didn't look as though anybody who would, could beat it, and those who possibly could, wouldn't.

Nobody could have blamed Shelby if he had sat back and rested on his and his crew's achievements. But Carroll & Co., including such foxy types as Ken Miles, have been around a good while. As we said, 1964 was strictly The Year of the Snake, but since the day when we walked into a small stall shop and helped put the finishing touches on totally unpainted Cobra #1, we've made a point of keeping an eye on the ever-growing snake pit.

So it was that at Sebring in 1964, even though we had other fish to fry (somewhat inauspiciously) we made a special effort to be around the Cobra quarters, having heard earlier that the ubiquitous Mr. Miles had shoehorned a monstrous Grand National type 427 C.I. Ford into one of the team cars. Sure enough, he had. He had also managed to mow down the only tree in five acres of open ground when the beast got away from him coming out of the Esses in an early practice session, and was busily pounding out the divots in the car when we made the scene. We decided this was no time for idle conversation and went elsewhere to attend to our own machinery.

During practice we made it a habit to glance nervously in the mirror when coming out onto a straight. If what we saw in the glass was blue with white stripes (or red) we moved over right now, due to the fact that whatever it was, was going to overtake right now. On one occasion we looked in the mirror, saw something blue with white stripes. We moved over and waited, and waited, and waited. We waited until we got about two thirds of the way down the straight. Suddenly there was this godawful explosive roar and our car picked itself up and set itself down at what felt like about three feet to the right. Something blue with white stripes was disappearing around the next turn. Ahrrgh!

The next question was: if he was going so bloody fast why did it take him so blasted long to get there? We found out the answer later. Every time Ken rounded a bend and headed down a straight the beast just sat and scratched until, eventually, it caught a bite, at which point everything was peachy keen until the next turn. So much for history; suffice it to say that the debuts of both cars, his and ours, were somewhat less than satisfactory. Tires available at the time and the buggy spring suspension of the older Cobra

Below right, the street version costs $7495 and is just about complete for that use. The competition version – as tested – costs $9500, but options needed for specific types of races can raise its cost a bit more. Directly below, the light-weight version of the powerful 427 used in the race cars has aluminum heads and other parts to make it some 80 lbs. lighter.

just weren't up to handling the 480-odd Percherons that lived under the hood. It also got more than somewhat warm in the cockpit.

The second appearance of this monster, widened, beefed, vented and with much bigger tires was Nassau '64. It still had buggy springs but not much else was the same. It went like the hammers and was only beaten, and just, by a lightweight GS Sting Ray in the hands of Roger Penske, and then only because some chassis bits broke up and Ken had to slow down.

We thought it still had to be only an exercise and something to keep the Shelby competition department busy in odd off moments. Surely Ol' Shel wasn't actually planning to build 100 of these brutes and turn them loose on the body politic? He for dadburned sure was! Fun is fun, but Shelby wasn't spending loot sending that prototype around to odd parts of the world just for the kicks of seeing Miles stuff his hoof into more horsepower than most people use in a lifetime. If the car was going to qualify for SCCA production racing he had to abide by the rules and build street machines. One full hundred of them.

By this time Miles and Jim Benavides, the project engineer, had decided that this could not be just a mod job on the older 289. It had to look like a Cobra I, but that was about all. So they started with a relatively clean sheet of paper and built a new car.

First, Shelby's comment that "it works" notwithstanding, that transverse leaf spring suspension had to go. Second, everything else had to be bigger and stronger to take the torque and the heft of the giant 427-inch powerplant, a unit that was designed to push 4,000 pounds of brick-like sedan at 170 mph. In fact everything had to go but one item—surprisingly enough the standard Salisbury rear-end used in the 289 was found to be more than sufficient.

Integral coil-shocks, as used on virtually all modern race cars, were selected both for compactness and ease of adjustment for chassis tuning. This, naturally, necessitated a complete redesign of the front and rear structures to carry the necessary upper A-arms that replaced the leaf springs which formerly located the upper ends of the hub carriers. The frame tubes went from three to four inches in diameter and were moved 22 inches apart, from the

The street 427 has two Ford-made four-barrels supplying fuel to the powerful engine. This setup produces almost as much torque at the bottom end as the competition version does.

PHOTOS: DARRYL NORENBERG

The racing 427 uses a single Holley with its centerpivot floats. It has a rating of 780 cubic feet per minute. Output from the 7000-cc monster is 485 horses with 480 lb. ft. torque.

original 17 as on the 289. The body design is that of the 1965 competition version and drops straight to the frame tubes instead of tucking under as before. To make room for the bigger bellhousing the pedal support box no longer straddles the frame, but is moved out with the widened tubes, thus offsetting the pedals to the left. This is disconcerting at first, but can be gotten used to in a day or so.

Bigger hubs carrying bigger, more widely spaced bearings, taper instead of roller, are used in the 427. Bigger, much stronger half-shafts are used and, though these are splined, the rear suspension geometry is such that only 0.020 of an inch of spline movement occurs in spite of more than six inches of wheel travel allowed. Designing the axle for spline travel would have been useless in any event, according to Miles, since the splines would bind up solidly under the brutal torque of the big-incher.

For the street machine, the suspension mountings are in rubber, and an option of bronze bushings is set up for the race car. In the street machine the engine is an out-of-the-box Grand National 427, fitted with two AFB Holley carburetors as used on the Mustang. These have fairly small primary venturis and big secondaries. Thus the street Cobra is fed gently at the outset and is, if anything, more docile than the 289 if one remembers that the first throttle stop is only a detent to remind you that the big secondary vents are about to open. On the race car the feeding is done by one great huge Holley 780 CFM as used on NASCAR machinery, sitting in a bathtub-like cold air box. Even this one is fairly docile at first, but it comes on much more rapidly. The carburetors on the dual set-up tend to flood or starve on very tight turns; the big single, with its concentric float, does not — most definitely not.

So it's docile at first. However, if one is in either of the two lower gears one is advised to make sure the rear end has caught a bite before one goes beyond that first throttle detent. This Hawg will lay great huge strips of rubber if treated disrespectfully at any speed in either first or second, and will even do it in third gear if it is the least bit out of shape. Treated with respect, the 427 is smooth and very deceptively fast in either top speed or acceleration.

Once upon a time, Aston Martin made a big thing out of going from zero to 100 and back to zero in less than 30 seconds. Ken Miles took one of the street 427 machines out and did zero to 100 to zero in 13.2 sec-

There can be few complaints about trunk space in the street version. It has lots of room.

Hidden in the upper/right corner of the race-car trunk is a pair of heavy-duty fuel pumps.

The street version, with its smaller-section tires, develops some oversteer when accelerating through a corner, but its manners are exceptionally mild for such a hot performer.

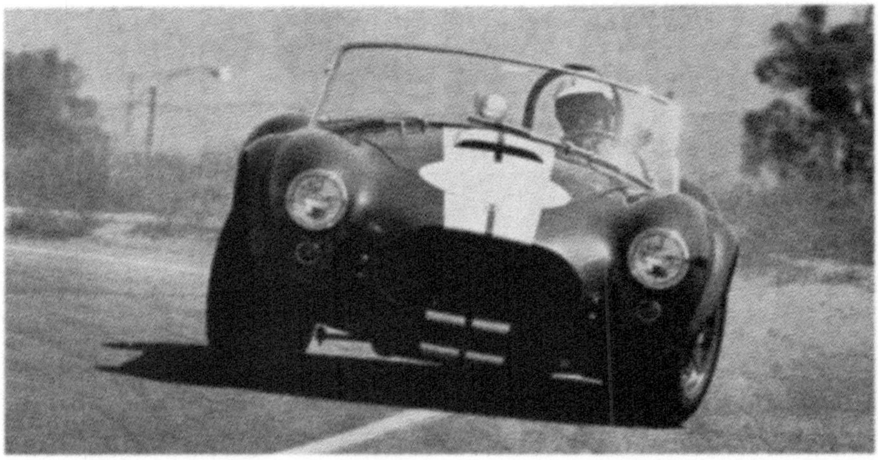

Directly above: Titus caught in the act! With low tire pressure, the race car's understeer used more road than he figured on. The photog was glad he'd used a mild telephoto lens, too.

427 COBRAS

onds. This was done on concrete. On asphalt things get a bit slicker. Due to wheelspin with the street tires (Goodyear Blue Dots), and the fact that the straight at Riverside had been used for a big drag meet wherein several cars had grunched on the line, the best we could get with the street machine was 110 mph in 13.2 seconds through the quarter. It was the first street machine we'd ever driven that would keep the tires lit up for the full quarter in every gear. The race car, with its "blueprinted" engine and huge rear tires, got a full second off of that, blasting through the trap at 6500 in third gear.

At one point we were entering a freeway and pausing at the entrance. Our passenger looked at the traffic and commented "Careful, the traffic is moving pretty fast — wasn't it!" That was how long it took to get up to and above the normal 70 mph flow of cars. To be very blunt about it, we do not advise those not extremely well-versed in the Cobra mystique to try the shenanigans conducted by Ken Miles in this car.

Looked at as a whole, the Cobra 427 is not a machine for general transportation. It is, rather the latest example, and very probably the pinnacle, of that vanishing breed, the pure sports roadster. Its water-tight integrity is on the minimal side, and it wouldn't be much fun to drive in most urban areas, particularly east of the Mississippi. It can be used for travel and there is considerably more room in both trunk and cockpit than in past versions, but best the weather be good and the spaces travelled be wide.

Our test car of the street version was bright red and we were thankful that in our area the law enforcement agencies are reasonably tolerant of such machinery, and that the exhaust system of the 427 is virtually silent except at full throttle. Painted red, this car is one that in some areas could get you a ticket while parked in a driveway. To sum up, our early misgivings to the contrary, the Cobra 427 rides comfortably and is bags of fun as long as one puts the brain in gear before putting the car in gear. One might get a bit wall-eyed keeping one eye on the road and the other on the rear view mirror, but that's been the name of the game for years, and a small price to pay for the privilege of driving a machine not seen on the road since the days of the Mercedes race cars of the 1930's.

We tested both the street version and the track machine. For this last, we will now turn you over to Jerry Titus:

Last Spring, Ken Miles debuted the competition 427 Cobra in the Riverside USRRC. Naturally, he had to run it as a Modified at that time. Aboard the Webster Two-Liter, we sat in his trunk for quite a few laps and had a good opportunity to watch the suspension work. It wasn't very impressive, but the thing stopped and accelerated like the best Modified on the track that day. A lot of careful development has taken place since then, and we again had the opportunity to be on the same track with a 427 — this time in the hands of Skip Scott — at the recent Continental Divide USRRC (See, Boss? We're out there evaluating other equipment when we're racing). This time the results *were* impressive. Scott was getting goodly amounts of power to the ground coming out of a corner, and quite a bit more bite under all conditions than the original prototype did.

Power is something the 427 has plenty of, yet the race-car version uses a single four-barrel Holley carburetor as opposed to the two quads on the street version. To reduce engine weight, cast aluminum heads, oilpan, high-riser manifold, and timing-chain cover are part of the competition package. This pares some 80 pounds from the stock 427. Torque is ample to bust the rear wheels loose from 2800 to 6500 rpm, so the gearbox ratios are fairly widespaced by race car standards.

Strong magnesium wheels of the pin-drive type, with knock-off hubs, are standard on the Cobra.

Fully-instrumented cockpit is simple and neat in the very best of performance-car tradition.

Editor Christy barrels the street version in an uphill turn on the fast Riverside circuit.

The street car is equipped with an electric fan for emergency use in hot-summer traffic.

The racing version has its oil cooler mounted in a rather vulnerable position under grille.

The shift handle is long but positive, with the throws very firm and very short. In the bottom three gears, delicate use of the throttle is required if you wish to continue heading in the direction desired . . . even with a machine that grosses (with fuel and driver aboard) in the neighborhood of 2700 pounds!

Obviously, the 427 Cobra is a brute. There's no other word for it. So the majority of effort in the race car version has been concentrated on helping the chassis cope with the performance. Huge Koni adjustable shocks are installed in both front and rear. Spring height is also adjustable via a screw-up collar that acts as a lower seat for the coils. CR Girling calipers up front and RR units in the rear clamp on 11-inch discs that are $\frac{9}{16}$-inch thick. The

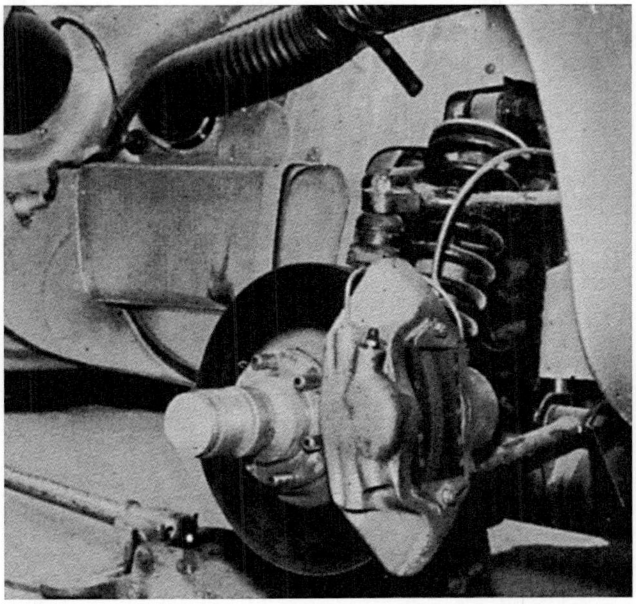

stopability of the heavy machine is strictly amazing. These are real anchors and the suspension works well in combination with them for maximum stopping stability; there's a lot more squat than nose-dive. Cooling-air ducts of fibreglass are provided for both front and rear discs.

CONTINUED ON PAGE 49

The street machine, top, has big 10.75-inch discs with unequal-arm suspension, coil/shocks, and front swaybar. Duct above brings fresh air to the cockpit. Notice the full undercoating.
The race car, above, is equipped with binders over 11½ inches in diameter and heavy-duty calipers to make it well able to handle the potential of the engine. Fiberglass duct brings in cool air.
Fully independent rear suspension is now the feature of the Cobra, at right, via the addition of upper A-arm. Armstrong coil/shocks are adjustable to both car-height and dampening setting required.

Cobra 427 Street and Competition Models.

Street: $7495 FOBLA
Iron engine with two Ford, four-barrel carbs.
HP — 425 @ 6000 rpm
Torque — 480 @ 3700 rpm
Curb weight: 2450 lbs.
Fuel Tank: 18 gal.
Wheel Size: 7.5 x 15 in. (front & rear)
Gearbox ratios:
1st — 2.23 to 1
2nd — 1.69 to 1
3rd — 1.29 to 1
4th — 1.0 to 1
Limited Slip Diff. — optional
Suspension: Large coil springs, tubular shock absorbers, front sway bar.
Brake Discs: 10.5-in. front
10.5-in. rear
Standing quarter mile with equipment as tested: 13.2 sec. @ 110 mph

Competition: $9500 dry sump
$9750 wet sump
Lightweight 427 with single 4bbl Holley carb.
HP — 485 hp @ 6500 rpm
Torque — 480 lb/ft @ 3500
Curb weight: 2350 lbs.
Fuel Tank: 42 gal.
Wheel Size: 9.5 x 15 in. (rear)
7.5 x 15 in. (front)
Gearbox ratios:
1st — 2.23 to 1
2nd — 1.54 to 1
3rd — 1.19 to 1
4th — 1.0 to 1
Limited Slip Diff. — standard
Suspension: Adjustable-height Armstrong coil/shock units, front & rear sway bars.
Brake Discs: 11.63-in. front
10.75-in. rear
Standing quarter mile with equipment as tested: 12.3 sec. @ 121 mph

THE KING AND HIS COURT:
The 427 Cobra and GT350

Carroll Shelby, producer of the world champion Cobra, holds court to formally introduce his Shelby American challengers for 1966 — the new 427 Cobra in the background, and the Shelby GT 350 in the foreground.

CARROLL SHELBY, creator of the world champion Cobra, has brought forth another challenger to international sports car producers in the form of the new 427 Cobra.

This sleek successor to the 1956 World Manufacturer's Champion 289 Cobra embodies the same quality and dependability that has made the name Cobra synonymous with success.

With a gross weight of 2150 pounds, powered by a 427 cubic inch Ford high performance engine producing 425 h.p., the new street Cobra enjoys nearly a 6:1 power-to-weight ratio.

New all - independent coil suspension and self-cooling disc brakes, coupled to the enormous driving power, enable the 427 Cobra to travel from a dead stop to 100 m.p.h. to a dead stop in a fraction over 14 seconds. In additional performance runs the 427 Cobra has turned 0-60 m.p.h. in 3.97 seconds, and 0-100 m.p.h. in 7.86 seconds.

The hand fabricated aluminum body of the 427 Cobra is made in England by A. C. Cars Ltd. The lightweight body shell is then fitted to the all new tubular chassis before being shipped to Shelby American in Los Angeles, Calif.

At Shelby American, the new Cobra body is mated with the powerful 427 Ford engine, a Ford four-speed synchromesh transmission and rear-end assembly. Prior to shipment to the Shelby American network of franchised dealers, the 427 is standardly equipped with $7^1/_2$ x 15 inch alloy wheels and 8.15 x 15 Goodyear special tires, wind-wings, sun visors and grill and rear-end bumper guards.

427 Cobra

The new 427 Cobra boasts a 427 cu. in. Ford High-Performance engine which produces 425 hp.

GT 350

The Shelby GT 350 features extra class with more sass for 1966.

Exterior changes of the 1966 Shelby GT 350 are readily apparent — the side ventilators have been replaced with quarter-windows, giving the rear window a "wrap-around" effect; air scoops have been added to the side body panels feeding air to the rear brakes; and the honey-comb grill has been replaced by a horizontal bar assembly and special emblem.

Inside, instrumentation has been changed to a round dial assembly with oil-pressure gauge in the dash cluster before the driver, and a new tachometer mounted atop the dashboard. A deep-dish steering wheel is another new interior attraction.

The things that haven't changed include the 306 h.p., 289 cubic inch Cobra engine with a 4V Holley carburetor atop an aluminum high-riser intake manifold, tuned header and exhaust system and various other Shelby American engine modifications.

The dependable "track-style" suspension system, with one inch front sway bar, anti-torque and stabilizer bars, that enabled the Shelby GT 350 to become a champion remains the same for 1966, along with the all-synchromeshed 4-speed transmission and standard competition-type safety belts.

New optional additions available soon will be a fold-down rear-seat and a high-performance automatic transmission. ■

The distinctive touch of Shelby American engineering and its design is pictured in this view of the rear-quarter-window and side air-scoop on the GT 350.

CAR and DRIVER ROAD TEST

COBRA 427

Not long ago, the Cobra 427 would have been the hot setup on any race track. Now it's a civilized street machine!

Several years ago, the manufacturers of a posh British grand touring car got a fair amount of mileage out of the claim that their vehicle could accelerate from 0-100 mph and brake to a complete stop in less than 25 seconds.

This was indeed an impressively brief period of time during which all that change of velocity happened, but automotive development has come a long way since then and today perhaps half a dozen production cars of one kind or another can perform on that level. What's more, there are several automobiles being produced in the United States that will break through that arbitrary 25 second barrier like the Germans through the Maginot Line. One is the 427 Sting Ray (see page 49); another, most certainly, is the new 427 Cobra from Shelby American.

Alright, you say, if 25 seconds from 0-100-0 isn't so hot anymore, what the hell is? Twenty seconds?

Forget twenty seconds.

How about 18 seconds?

Not too bad, but the Cobra can do better.

How much better, wise guy?

How about maybe 14.5 seconds? Get that, 14.5 seconds to accelerate to 100 miles an hour and then stop again. Until something better comes along, that may have to stand as some sort of high water mark in performance for cars that are readily available to the general public. That figure, mind you, is obtainable by the average Cobra driver with the regular 8.15 x 15 Goodyear Blue Dot street tires. Cobra test driver Ken Miles has done the job in as little as 13.8 seconds, and who knows how much improvement could be made with racing tires that would nullify some of the tremendous wheel spin?

The 427 Cobra does accelerate and decelerate at unbelievable rates, as the above figures should imply. What's more, it is a more civilized machine than the original 289 Cobra that brought the fabulous Shelby organization into being four years ago. It handles properly, thanks to a completely new all-independent suspension system that is traceable to the deft hand of Klaus Arning, the Ford Motor Company genius responsible for the impeccable handling of the Ford GT.

Everyone at Shelby is more than candid about admitting that the handling of the original Cobra was considerably less than optimum. In fact, *C/D* was once informed by a Shelby lieutenant that the old tubular AC chassis had considerably less torsional rigidity than the rail frame of a Model T! Coupled with this flexible frame was an antiquated suspension system, designed in the post-war years, that utilized leaf springs and lower wishbones. One staff member recalls a particularly painful day in southern California when he was outrun down a bumpy orange grove lane by an MG 1100. "There I was, with all that Cobra horsepower, and the rear wheels were bouncing and leaping around so badly that I could barely keep the beast on the road, much less catch up to the MG. It was terrible!"

He should try the same trip in the 427. The new frame, still fabricated at AC Cars in England—but to Shelby specifications—is as stiff as a Redwood trunk and permits the equally-new coil sprung suspension to operate at maximum efficiency. Arning has designed the same anti-dive and anti-squat characteristics into the 427 Cobra that he used so successfully on the Ford GT and they contribute immensely to the 0-100-0 times the car is able to record. Under heavy acceleration, the car tracks nicely for a machine with such power, and its braking manners are magnificent. The massive Girling discs haul the car down from 100 mph-plus speeds like you've suddenly run into a sand bank, and much of this is due to the suspension's anti-dive capability. The only defect we found in the Cobra's acceleration-deceleration performance was a nasty little habit of trying to dog track when the throttle is wide open. The car will break traction to speeds beyond 100 mph and imprudent applications of power will send the tail-end slewing sideways. This apparently is an inbred trait in all front-engine automobiles with power-to-weight ratios in the 6:1 range and no amount of suspension work can eliminate it entirely. Certainly wider-base racing tires will reduce the problem, but the fact remains that the Cobra 427 is not an automobile for novices.

Unlike the 427 Sting Ray, the Cobra has retained its identity as a raw-boned, wind-in-the-face sports car. While the Sting Ray is a completely civilized vehicle, available with everything from multiplex FM radio to air conditioning, the Cobra comes across the counter with the same kind of side curtains that English sports cars have carried since Sir Henry Seagrave first turned an ignition key. Another feature designed to delight the Purist is the hand-operated top, the erection of which may rank second only to folding up a road map for sheer, brain-addling complication. Some of the staff complained about these archaic fittings, claiming loudly that any automobile that lays claim to being contemporary should at least have roll-up windows and a power-operated top. Others defended the Cobra, arguing that its raw power, the great brakes and the advanced suspension create a vehicle with such unabashed appeal and excitement that the owner plain won't give a damn about creature comfort. He *might* object if he knew that at any moment an automobile could invade his chill, wind-buffeted world and blow his Cobra into the nearest ditch. But that just simply ain't going to happen. The driver of a 427 Cobra, at the moment this is written, has about as much fear of being passed by

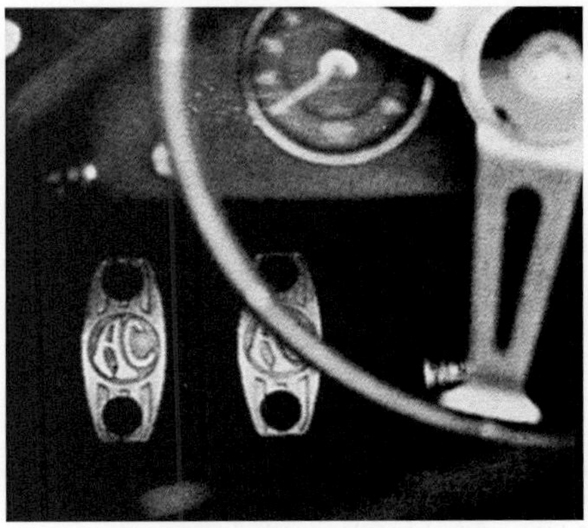

a herd of stampeding Water Buffalo as he does by a faster automobile, and *that* alone can make up for a lot of uncivilized traveling.

We tested the Cobra during the same Los Angeles heat wave that contributed to the tragic race riots and found the big car to be amazingly tractable. It refused to heat up, despite several hours of chuffing along on clogged freeways, and this was a welcome contrast to the old 289s, whose temperature gauges were inclined to rise clean off the scale at anything under maybe a sustained 80 miles an hour. The installation of a thick-core, 20-quart radiator, and a bigger grille opening aid greatly in keeping things cool, but the biggest safeguard against overheating is a small fan mounted ahead of the radiator that is thermostatically actuated whenever the water temperature reaches 70 degrees Centigrade.

Heat is a factor in the cockpit, however. With that great brute of a powerplant thumping away just inches ahead of the firewall, a substantial amount of heat is bound to penetrate even the best insulated flooring, and we found that temperatures around the feet were inclined to get awfully uncomfortable after a few hours running. Shelby American engineers are attempting to correct this problem with the use of more insulation, but we wonder if there simply just isn't too much heat to overcome.

Being about seven inches wider than the old 289, the 427 is a more comfortable car; about that there should be no question. The same basic Cobra layout remains essentially unchanged in the new car, includ-
(Text continued on page **49** *Specifications overleaf)*

427 COBRA

Manufacturer: Shelby American Inc.
6501 W. Imperial Highway
Los Angeles, California

Price as tested: $7000 (approx.)

ACCELERATION

Zero to	Seconds
30 mph	3.2
40 mph	3.6
50 mph	3.9
60 mph	4.3
70 mph	5.5
80 mph	6.2
90 mph	7.3
100 mph	8.8
Standing ¼-mile	118 mph in 12.2

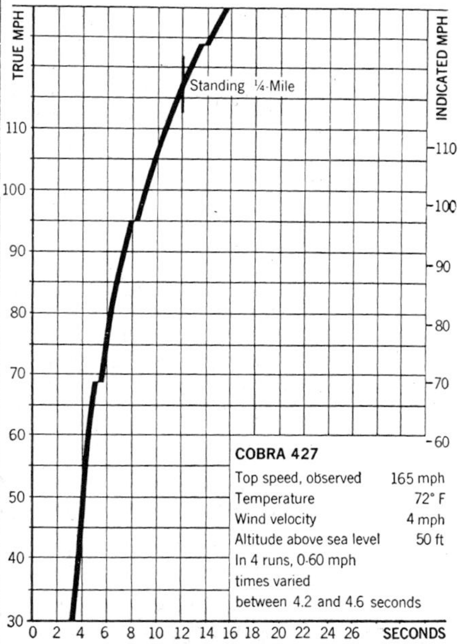

ENGINE

Water-cooled V-8, cast iron block, 5 main bearings

Bore x stroke.....4.24 x 3.78 in, 107 x 96 mm
Displacement...............427 cu in, 6998 cc
Compression ratio................10.4 to one
Carburetion..................Two 4-bbl Holley
Valve gear. Pushrod-operated overhead valves, mechanical lifters
Power (SAE)............485 bhp @ 6500 rpm
Torque...............480 lbs-ft @ 3500 rpm
Specific power output......1.14 bhp per cu in, 69.29 bhp per liter
Usable range of engine speeds..500–7000 rpm
Electrical system...12-volt, 70 amp-hr battery, 55A alternator
Fuel recommended..................Premium
Mileage............................9–12 mpg
Range on 18-gallon tank........162–216 miles

DRIVE TRAIN

Clutch..............11.5-inch single dry plate
Transmission..............4-speed, all synchro

Gear	Ratio	Over-all	mph/1000 rpm	Max mph
Rev	2.32	8.21	-9.86	-69
1st	2.32	8.21	9.86	69
2nd	1.69	5.98	13.54	95
3rd	1.29	4.57	17.71	124
4th	1.00	3.54	22.91	160

Final drive ratio....................3.54 to one

CHASSIS

Wheelbase...............................90 in
Track............................F 56, R 56 in
Length.................................156 in
Width...................................68 in
Height..................................49 in
Ground Clearance......................4.35 in
Dry weight............................2354 lbs
Curb weight...........................2529 lbs
Test weight...........................2890 lbs
Weight distribution front/rear.........48/52%
Pounds per bhp (test weight)............5.95
Suspension F: Ind., unequal-length wishbones with anti-dive and anti-squat, coil springs.
R: Ind., unequal-length wishbones with anti-dive and anti-squat, coil springs.
Brakes.....discs, 11.63-in front, 10.75-in rear, 580 sq in swept area
Steering......................Rack and pinion
Turns, lock to lock.........................2.5
Turning circle............................36 ft
Tires............8.15 x 15 Goodyear Blue Dot
Wheels......................7½ x 15 Cast alloy

CHECK LIST

ENGINE

Starting	Good
Response	Excellent
Noise	Good
Vibration	Good

DRIVE TRAIN

Clutch action	Excellent
Transmission linkage	Excellent
Synchromesh action	Excellent
Power-to-ground transmission	Good

BRAKES

Response	Excellent
Pedal pressure	Good
Fade resistance	Excellent
Smoothness	Excellent
Directional stability	Excellent

STEERING

Response	Good
Accuracy	Good
Feedback	Good
Road feel	Good

SUSPENSION

Harshness control	Good
Roll stiffness	Excellent
Tracking	Fair
Pitch control	Good
Shock damping	Excellent

CONTROLS

Location	Good
Relationship	Good
Small controls	Good

INTERIOR

Visibility	Excellent
Instrumentation	Good
Lighting	Good
Entry/exit	Good
Front seating comfort	Good
Front seating room	Good
Rear seating comfort	—
Rear seating room	—
Storage space	Fair
Wind noise	Fair
Road noise	Fair

WEATHER PROTECTION

Heater	Good
Defroster	Good
Ventilation	Poor
Weather sealing	Good
Windshield wiper action	Good

QUALITY CONTROLS

Materials, exterior	Good
Materials, interior	Good
Exterior finish	Good
Interior finish	Good
Hardware and trim	Good

GENERAL

Service accessibility	Excellent
Luggage space	Poor
Bumper protection	Poor
Exterior lighting	Good
Resistance to crosswinds	Good

ing those hilariously antique metal brake and clutch pedals with the art nouveau "AC" emblem, but everything is simply a bit roomier. The seats are deep, comfortable leather-covered buckets that would accommodate even Goose Tatum without great difficulty. We found that a six-footer sticks a good distance up over the doors, but he's still well protected by the windshield and any vulnerability he might feel is purely psychological. The steering wheel is perfectly positioned, though the shift lever comes out of the tunnel about three inches too far aft to be described as ideal. Although this causes no real upset, a reasonably tall driver will find that he has to bend his elbow as if he was getting ready to let fly with a bowling ball whenever he wants to engage first gear.

For a car that lays valid claim to being the fastest production machine in history, the 427 Cobra is amazingly simple. Its powerplant is the standard big Ford of the type that be purchased in any Galaxie at any local dealership. "It's a big cooking engine, with a rather peaky torque curve that produces a great horsepower reading for the customer and is ideal for flexible road driving. But we tune the competition 427s much differently, with a flatter curve," says Ken Miles.

One might expect a Cobra with an engine displacing 427 cubic inches to be an absolute beast on the street. It is utterly to the contrary, with a positively placid disposition at low speeds. This faked us out completely because we expected to find a machine with a vicious, bear-trap clutch and an engine that idled something like a Double-A fuel dragster. We found the 11.5 inch Ford clutch to be no more challenging than a normal domestic unit and the engine ticked off a 700-rpm idle with style and grace. In fact, the smoothness of the Cobra at low speeds completely belies its breathtaking performance, and only when the throttle is cracked does the driver realize the reservoir of power is practically a bottomless pit. A top speed of 165 mph is possible with the car's standard 3.54:1 final drive ratio, and that should be sufficient for travel on any thoroughfare except the Mulsanne Straight.

Like the engine, the transmission is a standard Ford four-speed unit that operates like all the other domestic all-synchros on the market today. Lest there be any misunderstanding, that means we love it.

The 427 Cobra is bulkier looking than its forerunner, and if anything, looks meaner. It utilizes the same wild fender flares that first appeared on the 289 racing versions, and based on pure subjectivity, we think the 427 Cobra is maybe the toughest looking car on the road.

Everybody knows wire wheels are out and Shelby American is supplying the 427 with Halibrand knock-off, magnesium wheels as standard equipment. They are specially fabricated for the car and are painted black, save for a polished rim and highlights around the air slots. It is possible that special Shelby-manufactured magnesium wheels will replace the Halibrands later on in the year, but the latter will remain as an option throughout the expected production run of about 200 automobiles.

Nearly a year has passed since the 427 Cobra was announced, and skeptics can still be found who will tell you there ain't no such thing as a production 427 Cobra. This is nonsense. There were at least 50 of the machines at the Shelby plant when we ran our test and more were arriving from England on a daily basis. The new Cobra is a reality and only approximately $7000 cash and the insatiable desire to own the fastest car in four counties stands between you and owning one. If you can scrape up the dough, we recommend that you take the plunge. Like they say, it'll never hurt you. Or at least it shouldn't. C/D

427 COBRAS

CONTINUED FROM PAGE 41

The cornering power of the competition car is excellent. Our test unit was provided with 15 x 8 Halibrand mag wheels up front and 15 x 9 rears. To these were mounted Goodyear stock-car treads in the R2 compound; 6.70 x 15 front, 9.00 x 15 rear. We understand that rim width and tire cross-section are soon to be increased further. There was only one problem with out test car; the optional bronze bushings were not installed at suspension pivot points and the normal rubber bushings allow considerable fore & aft movement of the suspension. Accelerating out of a corner, we found that the line would change two or three times as these rubbers loaded and unloaded. In a car of the Cobra's potential, this can be . . . ah, quite disconcerting. The word is that, selling for $9500 per copy, the full-race version actually represents a one-grand loss to the factory. Hence the attempt to recoup some of it by offering options like the bushings, a few more minor goodies, none of which were on the one we tested.

The cornering attitude of the car can be altered by suspension adjustment. As set, this one had strong understeer. It had *terrible* understeer until we discovered the front tires had only 26 pounds of air. These were pumped up to 36 pounds (the rears were left at 22 lbs.) and, by golly, it *would* turn a corner after all. We would have liked still less of it as, even with relatively light steering pressure, it took a lot of horsing and a very hot approach to get hung into a drift that would allow the power to be turned back on. Otherwise, throttle-application only produces more of the "let's-go-straight" attitude.

Straight-line performance is something out of this world. Playing around with a couple of quarter-mile runs, we hand-timed it at 12.3 seconds. Set up for dragging, it could probably better that considerably. Above 130 mph, the wind-buffetting lets you know it's aerodynamically a big brick, but the only instability is when slight bumps produce wheelspin. Top speed is reportedly in the 180-mph category. The equal of any big production sports car in a corner, it will get around even the tight circuits in impressive lap times. Miles has logged laps in the 1:35 category while testing at Riverside; a very solid Production-car record and a little over six seconds below the absolute track record. We would have liked to have tried for lap times with this unoptioned version, but four places on the track were dug up to put in waterlines. (Yessir, we said 'water.' There's even a square acre of green grass in the infield now). Even the unoptioned car has to be as quick as the best 289 Cobra and, since both are classed by SCCA in A-Production, it will soon make the latter somewhat obsolete. Its pretty hard to argue with 480 horses, which is roughly what the tuned 427 puts out with complete reliability, and with torque like a dump truck.

AC CARS

ENGLAND

Styled by Frua of Italy, the chassis by AC cars of England and powered by an American Ford, the AC represents an international production. The car offers excellent performance with a top speed in excess of 145 mph using the largest of two optional engines. A tubular frame is built around backbone tubes four inches in diameter to provide great strength and torsional rigidity without extreme weight. The aluminum body shell is not called upon to provide any stiffness to the structure. AC cars are available in three model designations which indicate the choice of engines. These include the AC 289, fitted with a Ford engine of that capacity; the AC 427 convertible, using the 427-cubic-inch Ford, and the Cobra 427 which is fitted with a high-performance 427-cubic-inch Ford by Shelby-American of Venice, California. Dimensions of the models are similar except that the wheelbase of the 427's is extended six inches to accommodate the larger engines.

ENGINES: AC 289: Ohv V8, 289 cu. in. 271 bhp @ 6000 rpm, 312 lb. ft. torque @ 3400 rpm, bore & stroke 4.00 x 2.87 in., compression ratio 10.5:1. One 4-barrel carburetor. **AC 427:** Ohv V8, 425 cu. in., 425 bhp @ 6000 rpm, 480 lb. ft. @ 3700 rpm, bore & stroke 4.24 x 3.79, compression ratio 11.1:1, two 4-barrel carburetors. **DRIVELINE:** 4-speed manual transmission, 3.54 rear axle ratio. **CHASSIS & SUSPENSION:** Steel tube chassis following racing car principles. Independent suspension front & rear wheels by wishbones, coil springs, and radius arms. **DIMENSIONS: AC 289:** wheelbase 90 in. **AC 427:** wheelbase 96 in., front track 54 in., rear track 53 in., overall length**(AC 289)** 166 in.; **(AC 427)** 175 in.; overall width 67 in.; height 51 in.; steering 2½ turns lock-to-lock; turning circle 34 feet; tire size 8:15 x 15.

Better known in the U.S. as the Ford Cobra, this beautiful sports car has wide appeal among the true auto enthusiasts. Equipped with the big Ford 427-cubic-inch engine, its performance is comparable to the best from Italy and Europe.

The steel tubular chassis follows racing-car principles. It has independent suspension, front and rear.

The driver sits well up with a good view across a long shapely bonnet; there is surprisingly little buffeting behind the screen.

MOTOR TESTED

The power game

A spectacularly fast car in a grand if dated tradition.

IF IT WERE not for the fact that the term Vintage is inextricably tied up with dates, one would be able to say with absolute truth that the A.C. 289 is the most splendid vintage sports car, a car that is characteristic of the image of the true masculine fun car. If you don't actually own one of the modern brand of sports car, whether it be a sophisticated Lotus or basic BMC, you may well think that they are all as stark and hairy as the A.C., but the 289 (the one-time Cobra) stands on its own as a car whose tremendous performance and roadholding is for ever a joy. Unfortunately the drawbacks are there too: it has a ride which many would consider poor; a primitive hood which you leave down unless it is actually raining when you are stationary; and little useful luggage space. To some these are the normal attributes of a sports car, part of the appeal of open style motoring.

The story of the sports A.C.s is an unusually long one with a concept which has remained virtually unchanged since 1953. John Tojeiro designed the chassis for the prototype shown in 1953; it used twin tubes with cross bracing, plus body tubes around which the aluminium panels were wrapped. Suspension at both ends was independent by transverse upper leaf springs and lower wishbones. The original Ace engine was the overhead cam six that had first been designed in the mid-1920s; the 2-litre Bristol engine was offered later in 1956. In 1961, still with basically the same chassis, a further option became available with the six-cylinder

Continued on the next page

Price: £2,400 plus £551 11s. 0d. equals £2,951 11s.

The whole shape of the car must be regarded as a classic.

Ford Zephyr unit in varieties of tune from Rudds of Worthing. It was in 1963 that Ford of America, in conjunction with Carroll Shelby, were looking for a race-worthy chassis for the 4.7-litre Fairlane engine; they chose the A.C., called it the Cobra, beefed up the chassis and went racing.

By 1965 the 380-odd bhp was not enough, and Shelby began to drop the 7-litre Ford engine into an even further revised chassis. The basic twin tube frame remained but the tubes grew bigger; still more important was the introduction of twin wishbone and coil-spring suspension at each corner replacing the transverse leaf springs. With Ford brain power behind Shelby, the installation was successful, but few were sold in Europe. With the termination of the A.C./Shelby contract and the sale of the name Cobra to Ford, A.C. were left with a more than adequate chassis for the 4.7-litre engine; the same chassis, lengthened, provides the basis for the Frua bodied 428 convertible, a superb car in the luxury tradition of A.C.

From 1953 to the present day, the two-seater body has remained basically similar with a revised nose coming with the Zephyr engine option and ever wider wing flares along with an ever wider track.

With 271 bhp in 21cwt of car, the 289 is the fastest car to 100 mph that we have tested, by just 0.1sec from the TVR Tuscan with the same engine. The unit is extremely tractable and it has an excellent gearbox. Although the cornering attitude of the car is largely dictated by the throttle at road speeds, the way the power is put through to the road, particularly on wet surfaces, is quite remarkable. At high speeds, around the 100 mph mark, it did not seem quite as stable as expected, but this was possibly due to worn rear wishbone rubbers on a two-year old demonstrator.

At £2,952, the 289 is an expensive toy if you are going to use it as a second car, but it is such fun to drive that it would probably get promoted to first car very quickly; it is not too impractical but a 15 mpg thirst and frequent servicing might put one off using it as a

Performance

Performance tests carried out by *Motor's* staff at the Motor Industry Research Association proving ground, Lindley.

Test Data: World copyright reserved; no unauthorized reproduction in whole or in part.

Conditions

Weather: Dry with little wind 0-8 m.p.h.
Temperature: 59°-70°F. Barometer 29.53 in. Hg.
Surface: Dry concrete and tarmacadam.
Fuel: Super premium 101 octane (RM) 5-star rating.

Maximum speeds

	m.p.h.
Mean of opposite runs	134.9
Best one-way run	135.3
3rd gear at 6,000 r.p.m.	98½
2nd gear at 6,000 r.p.m.	77½
1st gear at 6,000 r.p.m.	58½

Acceleration times

m.p.h.	sec.
0-30	2.5
0-40	3.4
0-50	4.4
0-60	5.6
0-70	7.2
0-80	9.0
0-90	11.3
0-100	13.7
0-110	17.9
0-120	22.8
Standing quarter mile	14.4

m.p.h.	Top sec.	3rd sec.
10-30	—	3.7
20-40	3.9	3.6
30-50	4.2	3.6
40-60	4.2	3.4
50-70	4.3	3.5
60-80	4.4	3.8
70-90	4.7	3.9
80-100	5.0	4.7
90-110	6.1	—
100-120	8.0	—

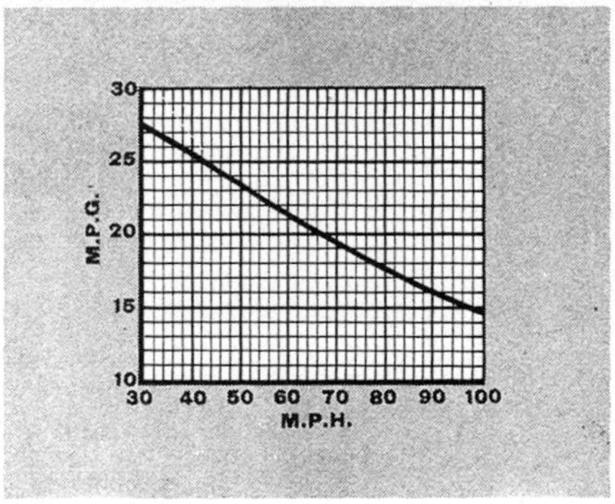

Fuel consumption

Touring (consumption midway between 30 m.p.h. and maximum less 5 per cent allowance for acceleration) . . . 16.3 m.p.g.
Overall . . . 15.2 m.p.g.
(=18.6 litres/100 km.)
Total test distance . . . 1,648 miles
Tank capacity (maker's figure) . . . 15 gal.

Brakes

Pedal pressure, deceleration and equivalent stopping distance from 30 m.p.h.

lb.	g	ft.
25	0.19	158
50	0.40	75
75	0.60	50
100	0.83	36
120	0.93	32
Handbrake	0.30	100

Fade test

20 stops at ½g deceleration at 1 min. intervals from a speed midway between 30 m.p.h. and maximum speed (= app. 85 m.p.h.)

	lb.
Pedal force at beginning	62
Pedal force at 10th stop	74
Pedal force at 20th stop	75

Steering

	ft.
Turning circle between kerbs: Left	33¾
Right	34¼
Turns of steering wheel from lock to lock	2.9
Steering wheel deflection for 50 ft. diameter circle	1.0 turns

Clutch

Free pedal movement . . . = ½ in.
Additional movement to disengage clutch completely . . . = 2¾ in.
Maximum pedal load . . . = 62 lb.

Speedometer

Indicated	30	40	50	60	70	80	90	100	110	120
True	31	41	52	61	71	80	89	98	108	117

Distance recorder . . . 4% slow

Weight

Kerb weight (unladen with fuel for approximately 50 miles) . . . 21.0 cwt.
Front/rear distribution . . . 50/50
Weight laden as tested . . . 24.7 cwt.

Hill climbing

At steady speed		lb./ton
Top	1 in 4.0	(Tapley 550)
3rd	1 in 3.2	(Tapley 670)
2nd	1 in 2.3	(Tapley 895)

Parkability

Gap needed to clear a 6 ft. wide obstruction parked in front:

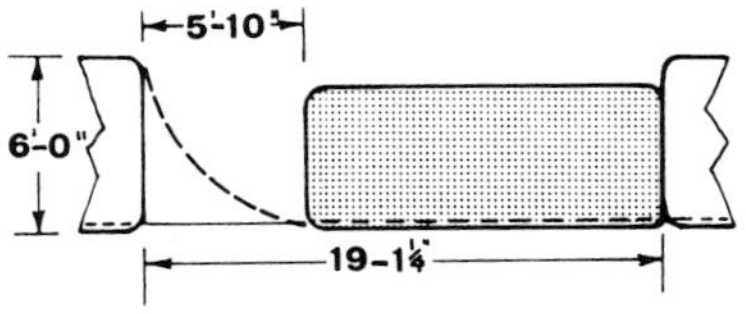

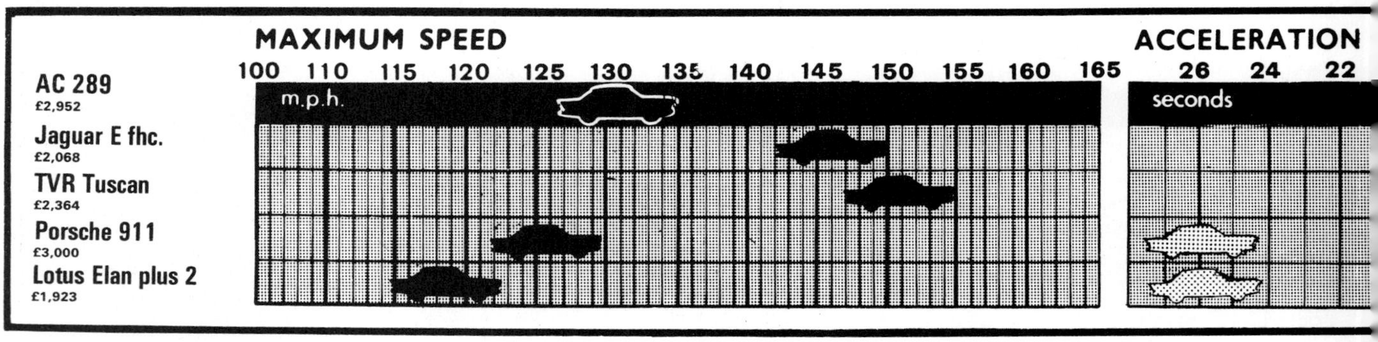

A touch of opposites to show how controllable the 289 is; it can be balanced at will on the throttle give or take a few feet of road space. Chrome bumper bar does not detract in any way from the purposeful aggressive appearance.

20,000-mile-a-year car. As lovers of controllable high performance we soon became attached to the 289 particularly with the roof down to reduce engine noise and hood flap. As we said earlier it has this splendid Vintage character which is not everyone's cup of tea.

Performance and economy

There are two ways of starting the big V-8; you can either pump twice on the accelerator or pull the choke out. The engine seems quite happy to idle straight away at less than 500 rpm and pulls without hesitation long before the temperature gauge moves off its bottom stop after some two miles. There is some tappet noise, mainly audible when the hood is up, but otherwise the engine is fairly unobtrusive. In normal road use you won't use more than 4,000 rpm and with the hood down you never hear the exhaust unless you accelerate past a wall, when the tremendous blast of sound conjures up all sorts of racing memories; if you wind on beyond 5,000 rpm towards the maximum around 7,200 rpm on this seemingly unburstable engine it becomes more frenzied but never alarmingly so. During our acceleration tests we used up to **6,500 rpm, reaching 60 mph in just 5.6 sec (still in first gear),** 80 mph in 9.0 sec and flashing past the quarter mile at over 100 mph—almost dragster levels.

If the sheer power is impressive so too is the smooth torque delivered throughout the range from under 1,000 rpm. From 20–90 mph each 10 mph increase only needs between 2.1 and 2.5 sec; 0–100 mph using only third gear takes less than 20 sec.

For our maximum speed we had to retire to foreign parts; the **mean 134.9 mph represented 6,300 rpm which was ear shattering with the hood up. With no oil cooler the oil temperature rose rapidly during these runs from its normal 80–100 deg C to over** 140 deg C, and it might be an idea to fit a thermostatically controlled cooler if really fast work is contemplated. Fuel consumption over that 640 mile trip worked out at 16.3 mpg with a lot of cruising in the 80–100 mph range, but the previous 900

Continued on the next page

A slightly cluttered dashboard provides all the necessary information; the gearlever is offset to the left but still falls to hand easily. Tall people found splayed knees fouled the handbrake.

Cast pedal pads carry the A.C. insignia and rubber buttons top and bottom stop the foot slipping off. Steering wheel has its wooden rim reinforced with aluminium. Rear view mirror is unframed.

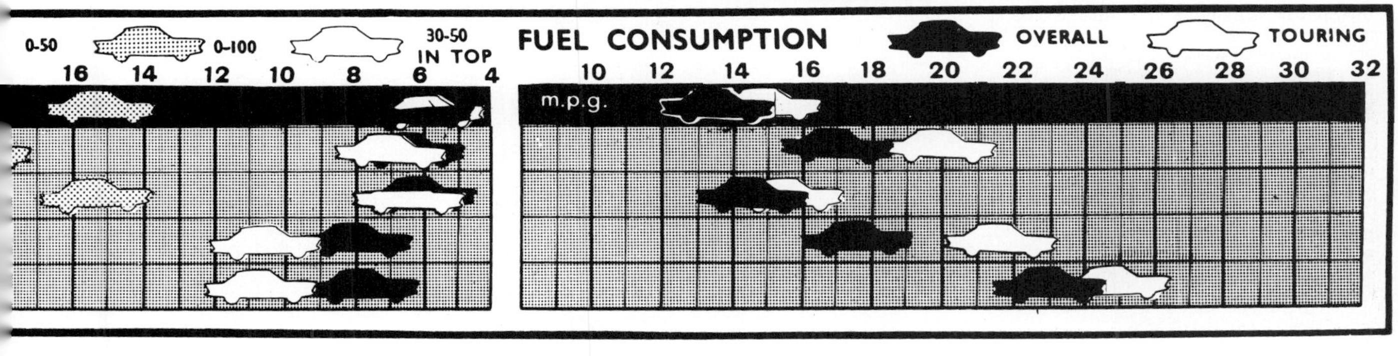

Rear bumper bar obstructs the boot handle. Brake cooling air outlets are behind the front wheel arch. The hood looks well tensioned but gets noisy around 70 mph. Side screens are removable and fit in a pocket behind the seats. Visibility is good with the hood up.

A.C. 289 Sports *continued*

miles in this country with a lot of town use and a fair amount of exhilarating blasting through the countryside only returned 14.5 mpg. You are unlikely ever to get better than about 17 mpg which gives a usable range on the 15-gallon tank of about 220 miles. Five star fuel is recommended but there was no pinking on four-star.

Transmission

When you have such tremendous torque a close ratio box tends to pass unnoticed until you come to do the standing start accelerations; for most manoeuvres out of town you could comfortably stay in top gear, and we frequently did when pottering through towns at 25 mph, but the gearbox has such excellent ratios and the short stubby gear lever is so pleasant to slide around that we all often used the box as if the car were a Mini with "...nothing below 3,500 rpm, old boy". The pedals are well placed for heel-and-toeing and there is room for the left foot beside the clutch, resting on a piece of floor which gets uncomfortably hot at times. The clutch itself grips smoothly and well but its long travel more or less dictates the seating position, and it is certainly the heaviest we have met for a very long time. With the seat adjusted for comfortable clutchwork the gearlever becomes well placed too; it requires very little practice to make smooth "family saloon" changes.

Since our test model was produced (since chassis No. 6120, in fact) these cars have been fitted with a 3.31 final drive instead of 3.54. This raises the gearing from 21.5 mph per 1,000 rpm to 23.0, and might add a few mph to maximum speed. Both final drives are Salisbury 4HU with limited-slip differential which contribute a lot to the confidence with which one can blast off past a line of cars for a quick overtaking burst in the wet; without it, life would be difficult. On standing starts we were able to get just as much wheelspin as we needed by controlled throttle use; we *could* have sat there with the accelerator on the floor and the wheels spinning helplessly, but with the clutch dropped in at about 2,500 rpm and the accelerator floored only when the speed of the car had caught up the rear wheels, we established our best times.

Measuring the Tapley pull in second gear sent the drum revolving to a record high, which worked out at the ability to climb a 1-in-2.3 hill. Small wonder then that the 289 romped away from rest on the 1-in-3.

Apart from a little gear whine in the indirects, and the occasional clonk from the limited-slip differential in parking manoeuvres the transmission was unobtrusive.

Handling and brakes

It takes a little time to be able to use the 289 chassis and power to best advantage; you start by coming into the corners too slowly and then blasting out too quickly. The tail is unlikely to unstick even so, but there are smoother ways of getting round. It pays to get the braking done beforehand, particularly as the rear brakes lock first, then go round the corner under more or less constant power until the exit is in sight. You can then use more throttle keeping the tail just in check, or let it come out a fraction lined up ready to squirt down the next straight (we tried all this out on a closed track). If the tail comes out a little it is easy to keep under control with a little less throttle and a bit of correction; but if it comes out a lot, and we had it up to an outrageous 60 deg attitude angle in the course of our experiments, it is difficult to bring back cleanly without a self steering lurch as the car comes back on the right keel. However, we still caught it, emphasizing what an extremely controllable car the A.C. is; with all that power it is quite a tribute that it handles so well and of course at very high cornering powers.

On wet roads we learnt to be careful having at one point lost the tail while pulling out to overtake. It came back as soon as power was eased but we developed the habit of using one gear higher than we would on dry roads; this reduced torque suitably without sacrificing too much of the phenomenal acceleration. We discovered shortly after the first excitement that the rear tyres had reached the throwaway level of $1\frac{1}{2}$–2 mm; with new tyres, Dunlop SP41HRs as before, wet road holding was rather better but staying a gear high was still a good maxim; with all that power you develop motorcycle sensitivity to road conditions.

This is all helped by the excellent steering which gives just the right kind of feel for safe fast driving. There is kickback on bumps

Safety check list

Steering assembly	
Steering box position	In front of axle line
Steering column collapsible	Yes with 2 UJs
Steering wheel boss padded	No
Steering wheel dished	No
Instrument panel	
Projecting switches	Yes
Sharp cowls	No
Padding	Covered facia with rounded metal top
Windscreen and visibility	
Screen type	Laminated
Pillars padded	No
Standard driving mirrors	Interior only
Interior mirror framed	No
Interior mirror collapsible	No
Sun visors	None
Seats and harness	
Attachment to floor	Sliding runners fastened to floor
Do they tip forward?	No
Head rest attachment points	None
Safety harness	Attachment points provided
Doors	
Projecting handles	No
Anti-burst latches	No
Child-proof locks	No rear doors

The boot takes a reasonable 3.9 cu-ft., of our luggage. The width is 37" with longest diagonal $49\frac{1}{2}$". The toolkit is usefully comprehensive. Copper ended hammer is used for the knock-off wire wheels. Grease gun it used every 2,000 miles.

through the rack-and-pinion mechanism, but it is nicely geared and surprisingly light at parking speeds. At high speed very little movement is required and you are hardly conscious of doing more than vary the pressure on the steering wheel to get round quite a noticeable 100 mph corner. This extreme sensitivity at speeds around the 100 mph mark leaves no room for heavy hands or feet; we were also a little peturbed at the tendency to veer to the right when power was eased after a full throttle blast at high speed. We hadn't really noticed this on the road, but during our acceleration tests on MIRA's narrow horizontal straight it was very evident, and it meant that the third to top change at over 100 mph had to be rather smoother than our normal rushed changes.

This directional variation is due to rear wheel steering; we suspect that newer cars or cars that haven't had a long spell as demonstrators, would not exhibit this feature, since it is almost certainly a sign of worn wishbone bushes; the wishbones are fairly wide-based but not comparable with modern racing designs which have to cope with the same power. Lifting off from steady speeds rather than full throttle makes less difference, and the car feels impressively stable and firm on the road even in strong side winds.

Looking back through the road tests of previous A.C. sports cars, we have recorded in each one that the rear wheels lock first which rather limits the maximum stable braking. With the 289 this is still true, and we could only get 0.93g using a high 120lb pedal pressure. On the road the brakes feel heavy (no servo is used) but firm and fade-free, but during our fade test stopping from about 82½ mph there was a steady pressure rise from 62lb to 75lb, although no loss of feel; the water splash produced a slight pressure rise for the first two stops but we only dared to make one trip as the water gushed in through the floor. The handbrake, which has a separate caliper, held the car on a 1-in-4 hill but did not have enough adjustment to hold it nose down on a 1-in-3, although the potential was there.

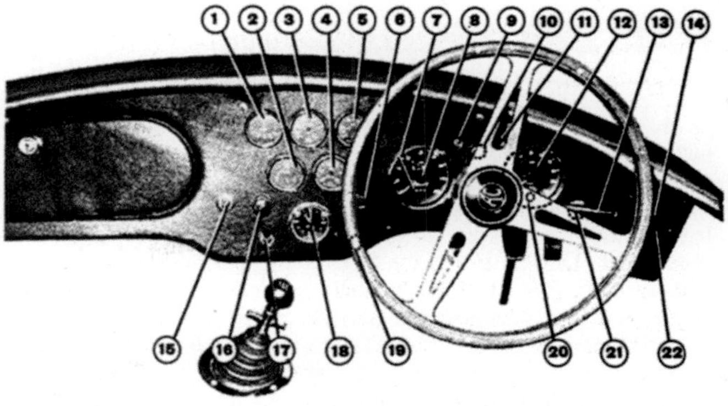

1, ammeter. 2, fuel gauge. 3, water temperature gauge. 4, oil pressure gauge. 5, oil temperature gauge. 6, two-speed wiper. 7, trip and total mileage recorders. 8, speedometer. 9, left indicator tell-tale. 10, dynamo warning light. 11, right indicator tell-tale. 12, rev counter. 13, indicator/horn stalk. 14, flasher/dipswitch. 15, two speed heater blower. 16, panel light rheostat. 17, side/headlight switch. 18, clock. 19, ignition/starter key. 20, main beam tell-tale. 21, choke. 22, washer button. Right hand fresh air control is obscured behind wheel rim at 4 o'clock.

Comfort and controls

With the track increased from 4ft. 2in. in 1953 to 4ft. 6in. now, the springs stiffened and the dampers made firmer there is very little roll; the ride is firm round town, though with little

Continued on the next page

Specification

Engine

Cylinders	V-8
Bore and stroke	101.6mm x 72.9mm
Cubic capacity	4.727 c.c.
Valves	Pushrod ohv
Compression ratio	11.0:1
Carburetter	Ford 4-barrel
Fuel pump	Ford mechanical
Oil filter	Ford by-pass
Max. power (gross)	271 bhp at 6,000 rpm
Max. torque (gross)	312 lb.ft. at 3,400 rpm

Transmission

Clutch	10½in. s.d.p.
Top gear (s/m)	1.00
3rd gear (s/m)	1.31
2nd gear (s/m)	1.66
1st gear (s/m)	2.20
Reverse	2.20
Final drive	Hypoid bevel 3.54—Salisbury limited slip
Mph at 1,000 rpm in:—	
Top gear	21.5
3rd gear	16.4
2nd gear	13.0
1st gear	9.8

Chassis

Construction	Twin tube frame with hoops and cross bracing—aluminium body

Brakes

Type	Girling disc brakes
Dimensions	$11\frac{11}{16}$in. dia. front, 10¾in. dia. rear
Friction areas:	
Front	27.6 sq. in. of lining operating on 249.4 sq. in. swept area of disc
Rear	21.1 sq. in. of lining operating on 195.8 sq. in. swept area of disc.

Suspension and steering

Front	Independent; wishbones and coil springs
Rear	Independent; wishbones and coil springs
Shock absorbers:	
Front	Armstrong telescopic
Rear	Armstrong telescopic
Steering gear	Rack and pinion
Tyres	Dunlop SP41HR 185-15
Rim size	6½L-15

Coachwork and equipment

Starting handle	None
Jack	Screw type
Jacking points	Chassis tubes and wishbones
Battery	12-volt negative earth, 45 amp hrs capacity
Number of electrical fuses.	4
Indicators	Self-cancelling flashers
Screen wipers	Lucas two-speed
Screen washers	Lucas electric
Sun visors	None
Locks:	
With ignition key	ignition/starter only
With other keys	Boot and glove locker
Interior heater	Recirculatory with two-speed blower
Extras	Detachable hard top chrome wheels, radio
Upholstery	Leather
Floor covering	Carpet
Alternative body styles	None

Maintenance

Sump	6½ pints SAE 30
Gearbox	2½ pints SAE 80 or 90EP
Rear axle	2½ pints Shell 6721A
Steering gear	SAE 140 hypoid
Cooling system	24 pints (2 drain taps)
Chassis lubrication	Every 2,000 miles to 8 points
Minimum service interval	2,000 miles
Ignition timing	10° b.t.d.c.
Contact breaker gap	0.018—0.021 in.
Sparking plug gap	0.028—0.032 in.
Sparking plug type	Autolite BF 32
Tappet clearances (hot)	Inlet 0.020in.; Exhaust 0.020in.
Valve timing:	
inlet opens	46° b.t.d.c. } at 0.008 clearance
inlet closes	84° a.b.d.c. } at 0.008 clearance
exhaust opens	94° b.b.d.c. } at 0.010 clearance
Exhaust closes	36° a.t.d.c. } at 0.010 clearance
Front wheel toe-in	$\frac{1}{16}$in.
Camber angle	Front 0°, Rear 2° negative
Castor angle	3°
King pin inclination	11°
Tyre pressures:	
Front	26 psi
Rear	30 psi

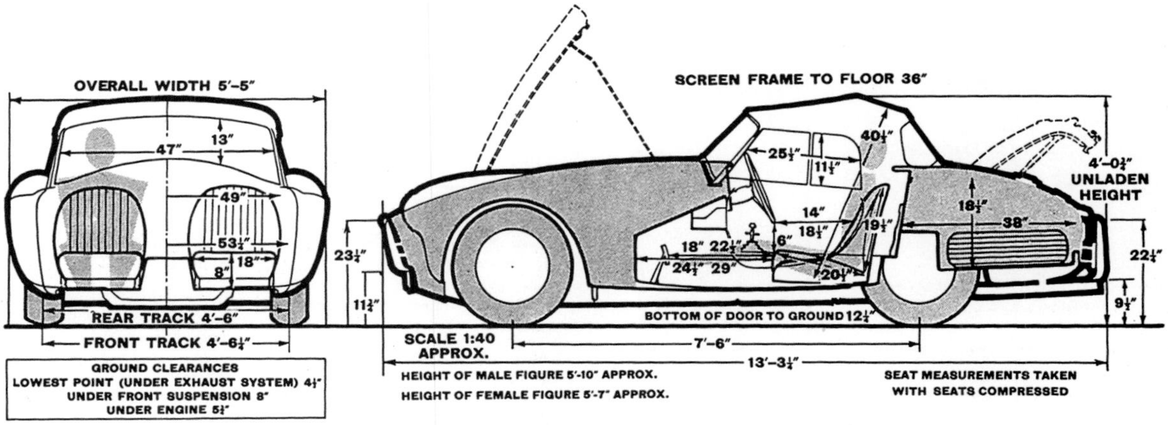

or no radial thump, and the car follows the road contours faithfully. At speed on main roads it is comfortable. On poorly filled-in roadmenders' holes you get the odd rattle from the rear suspension, but otherwise the car feels rigid and free from scuttle shake on surfaces which might be expected to produce it. Out of town, the ride remains firm but gives that reassuring feel of the road that is essential in fast cars; you get jerked in the slightly bouncy seats on sharp bumps, but on wider spaced undulations it just follows the road with no wallowing and no discomfort for that type of car.

On the Belgian pave the car behaved fairly well; on good sections it would maintain 80 mph or so with only a loud rumble to show you were on a different surface, but on worse sections with potholes the speed was dropped to 50 mph for fear of using up all the rather small ground clearance. On typical French secondary roads with steep cambers and fairly ridged surfaces the wheels stayed on the road but there appeared to be some bump steering which occasionally spoilt the line into corners.

The driver's seat seemed to be a little more comfortable than the passenger's which had less lumbar support; both backrests were too upright but no one found them uncomfortable and they have good side support. The pedals are slightly offset with respect to the wheel, but the seat is angled so that you don't get the spinal twist of some offset layouts. The pedal pads are metal with rubber inserts and pivot through a small angle range to adapt to the angle of the foot, a pleasing feature of A.C.s for some time; those with shoes of size 9 or so occasionally jammed their clutch foot on the toe-board—a piece of untrimmed glass fibre; there is an adjustable steering column.

The soft hood doesn't appear to have changed in design since the first of this line; it is held on top of the screen by two pins in the centre which slide into a channel, and at the edges by over-centre clips. Press studs hold it down round the base of the hood, and there is a rubber sealing strip inside; tensioning is provided by a hoop hinged to the main arch.

Insurance

A group rating	7
Lloyd's	On application

1, starter solenoid. 2, battery. 3, radiator filler. 4, washer reservoir. 5, distributor. 6, oil filler. 7, coil. 8, dipstick.

MAKE: A.C. MODEL: 289 Sports MAKERS: A.C. Cars Ltd., Thames Ditton, Surrey, England.

The fabric and the side screens fit into the bag behind the seats. It takes about two minutes for one practised person to put up, but two could probably halve the time. Once up it is rather primitive for this day and age; you get leaks coming in round the side screens, and the noise is rather less bearable than the lesser buffetting when the hood is down. It is with the hood down that we all had our best memories of A.C. motoring; it is quite acceptable up to 110 mph or so and only at 120 mph does it get really uncomfortable.

Visibility is unusually good with the soft top; the screen pillars are thin and there is plenty of clear sheet in the fabric. With the speed limit, we found the lights good enough in this country where we either knew the road or were guided by Cats-eye studs, but in France, where neither condition existed, we felt that they could have penetrated further; as it was they limited the speed to around 80 mph.

With the hood up there are few draughts until you move the sliding sidescreen forward, which is necessary in traffic when the interior can get a bit warm. On the move you can get cold air in by ram effect at foot level; this is controlled by knobs at either end of the facia. For hot air you have to use the two-speed fan since the system is a recirculatory one; there are flaps at foot level and the screen outlets demist the screen quite quickly.

Fittings and furniture

There is a splendid number of instruments on display, but their layout is hardly ideal. The wheel spokes obscure the lower edges of the speedometer and rev-counter so that 30 mph is largely guesswork and taller people with the seat right back found the oil pressure gauge also hidden. We liked the horn operation, a bar in the back of the indicator stalk, but didn't like the flasher/dipswitch stalk on the facia with its two-plane movement. There is no padding on the painted metal facia top and its flat surface reflects overhead trees on to the screen.

There is a glove locker on the facia and you can get a brief-case behind the seats, but the rest of the luggage has to go in the boot on top of the spare wheel. In fact the boot took 3.9 cu.ft. of our luggage, or rather more than either the Mini or the long tailed Hornet. Maps can fit in either door pocket.

Carpets are fitted to the floor and transmission tunnel and the seats are upholstered in leather; the door locks at the rear of the panel are operated by a simple leather pull-strap and you have to open the sidescreen and pull the strap to undo the door from the outside. No safety belts were fitted to our test car but anchorage points are provided behind the seats.

Servicing and accessibility

The bonnet is retained by two top catches at the rear and hinges at the front end; it is easy to lift and is propped open by a single rod with a rubber tipped end. There must have been a lot of room around the original six-cylinder engine because all the parts on the V8 are easily accessible.

The 289 needs servicing every 2,000 miles for which the handbook gives reasonable instructions to the owner mechanic. There aren't many dealers in the country, but A.C. retain a personal contact with all owners and can arrange the nearest suitable agent; there are in fact fewer than 30 of the coil sprung V8 engined cars in this country and only 60 leaf sprung ones, since the vast majority went to America.

As can be seen from the boot picture a comprehensive toolkit is provided. The jack has to fit under the chassis tubes or under the wishbones, but with a flat tyre there is little room to get it under; you need to spread the tonneau cover out on the ground, lie down and grovel underneath.

M

Maintenance summary

Every 2,000 miles: change engine oil, grease drive shaft joints, grease drive shaft splines, grease prop. shaft joints, check master cylinder reservoirs.

Every 4,000 miles: change differential oil, change gearbox oil, grease steering pinion bearing, oil distributor, oil accelerator shaft, grease handbrake nipples, oil carburetter linkages.

Every 10,000 miles: inspect front and rear hub bearings and grease sparingly, oil dynamo front bearing, grease steering gaiter ends. Occasionally oil wiper arm pivots, grease battery terminals, oil locks and hinges.

THE AC ACE was originally designed to take 80 bhp or so from the old AC 2-litre engine. I have just been proving that what is recognizably the same vehicle can absorb over 500 bhp from a "hot" 7-litre Ford V8 engine, and that in this guise it is a superb touring car!

JOHN BOLSTER tries

A 7-litre AC Cobra

This car was one of a pair built for a wealthy American driver. The other one was purely a track car, but the present machine has all the road equipment of a standard 4.7-litre Cobra. However, the 7-litre engine has Shelby aluminium racing cylinder heads and it also has a "full race" bottom end, including special connecting rods at £30 each. It differs from the 550 bhp racing unit in having a gentler camshaft for road use, and in fact its characteristics are ideal, for it will creep through London traffic in top gear or, with a bellow of rage, swing the rev counter needle past 6000 rpm in the same gear in a few seconds—and that means 144 mph on the 3.54 to 1 rear end which is at present installed.

With all that torque available, normal tyres would be quite incapable of transmitting the power to the road. However, this is one of 50 special Cobras which were built to give body clearance for immense racing tyres on wide light-alloy wheels. These are almost unbelievably effective, and all the power can be used all the time, except from a standing start.

The present owner, John Woolfe, has had the car converted to right-hand drive. This is not supposed to be possible with the 7-litre engine, owing to the position of the starter motor, but some modification of the bulkhead and pedals has allowed the standard British steering gear to be adopted. An effective hood with detachable sticks came with the car, but to enjoy it to the full I drove the machine in open form.

Driven slowly, this Cobra seems a real heap! The huge tyres make the steering feel sloppy and the rear end kicks about over bumps, with apparently no directional stability at all. When one releases a few hundred horses, the car instantly comes alive. At 150 mph it runs absolutely straight and true with no correction from the driver, who can relax with one hand on the wheel. In spite of its short wheelbase and wide track it is well balanced on corners, accepting a great deal of power when leaving a bend without getting sideways. On the wide open spaces of Snetterton I was surprised at the cornering power and controllability.

The acceleration is simply tremendous, and bears no relationship to that of any other road-equipped car. It is not only that the Cobra is light and has 500 bhp but, even more important, all that power can be transmitted to the road. Any sports car that one happened to encounter could be

overtaken without bothering to change down.

My first attempt at the standing quarter-mile produced a time of 12.8 s which, with a little practice, I reduced to 12.4 s. This was achieved by using only moderate wheelspin at the start but keeping just appreciable wheelspin going for a considerable distance. It was found best to change up relatively early, employing no more than 6000 rpm on any gear, and I did not snatch the changes with an open throttle. The clutch is perfectly happy to handle all that torque and the gearbox gives light, fast changes. For my fastest standing quarter-mile I used top gear for quite a distance towards the end of the run. From a standing start, 100 mph comes up in 9.8 s, using the three lower gears.

With the special racing bottom end, the engine is safe up to 7000 rpm, but I am too much of a mechanical purist to let an engine with nearly a litre per cylinder run that fast. With a really wild camshaft, it might pay to use such extreme revs, but with a road camshaft there seems no reason to do so. The final drive ratio fitted suits the average circuit very well and gives an easy 150 mph without over-revving. With a higher gear (lower numerical ratio) and a hard top on the body, extremely high speeds would certainly be possible, but the present gear ratio is a very good compromise for normal use.

The car takes a bit of stopping from high speeds. The brakes are there all right, and they do not fade, but a strong right leg is required for emergency retardation. However, when a corner rushed towards me and I realized that I had left it a bit late, my fear lent extra strength to my muscles and I did not enter the *décor*. A lady driver, for whom this car is otherwise perfectly suited, would want a bit more servo assistance on the pedal.

No better town and shopping car could be imagined. It is immaterial whether one uses first gear or third for starting from the traffic lights, as in either case the rest of the traffic just melts out of sight. The engine idles absurdly slowly and completely evenly, the typical V8 exhaust note smoothing out into quite a high-pitched scream as the big engine attains full revs. Apart from the exhaust, the engine is remarkably quiet mechanically and the gearbox and final drive are quite inaudible. At reasonably high speeds the ride is very comfortable, and the seats give good location. The handling is far better than that of other Cobras I have driven, the suspension obviously having been set up correctly for the enormous racing tyres. Unfortunately the stupendous performance must be paid for at the rate of 10 mpg or worse.

John Woolfe is very fond of his Cobra, as he has good reason to be. However, he is deeply involved in other racing projects and may find little time to make full use of this delightful machine. Anybody wishing to own a Lamborghini-eater might do worse than try tempting him with a cheque book.

QUINTS ON COBRAS

By ROMSEY QUINTS

Walk around it . . . don't lean on the alloy body panels . . . treat it with respect. It's one of the true bears of the sports car set of the sixties.

Romsey quivers to the bark of Detroit's TCH and discovers the bite of the Cobra is something else

"IT'S got some stuff," said Ian Geoghegan. "She's a bit frightening when you give her the big razzamoola."

"Oh?" says I, picking my teeth, like. I switch it on. Noise. Somewhere down in the bowels of the earth the subterranean trolls are playing tenpins. Select first. A hernia clutch, by God. Not so much drop the clutch as relax pressure from toe to ankle to calf muscle to thigh. Underpants tight. Head out the gate, calling back to Geoghegan, "Frightening for me I would not argue — but you!" I was about to find second when I heard his call back down the road . . .

"No, no . . . frightens me-e-e-e-eee . . .!"

Anything that can frighten Ol' Garbage, who spends a considerable amount of his Mustang time looking out the side windows to see where he is actually going, must be fairly horrific in demeanor. It was. Fred, I have just spent time with a big red AC Cobra, a neanderthal, slam-bam-thank-you-ma'am refugee from the racoon coats, a street dragster, a Detroit MGTC.

The AC Cobra was born of the Let's Just Drop In school (it doesn't go, so Let's Just Drop In some cubic inches). It was one of the first matings of a British chassis and an American V8, giving birth to a whole series of what Fred Markone is now calling Handy Tools but which I call Cardiac Cars. As you may gather, I am not one of the school that considers Pontiac GTOs as slightly warmed-over Valiant V8s.

The Big Red Thing was privately imported for somewhere around $10,000 as a road car, and had done only a few thousand miles when it arrived to join the rest of the weekend droolers on the lush yard of Geoghegan's Sporty Cars on Parramatta Road, Haberfield, on Sydney's street-of-car-yards. Apart from slightly puckered upholstery on the driver's seat, it was virgin ground, almost with the feathers still on the Dunlop SP41s.

Now, I am not exactly a beardless youth when it comes to time machines like this one, having driven a TVR-Griffith imported by Peter Owen in which I only once found the bottom of the accelerator travel before it was sold to go draggin', which I consider a just fate for it. But that had only a stock 289 kewbick incher in it, not the 271 bhp stoker in the Big Red Thing. If you pare about 8 cwt offa a Falcon GT and add another 40 bhp, you'll begin to see how this thing goes. Frinstance, if you smack it really hard after going from second to third it starts wanting to go sideways. Again. Still?

Walk around it. Don't lean on the alloy bodywork, if you don't mind. Cutaway doors, real honest cable door pulls, not this modern chromed flip-top junk, meshed air vents either side of the radiator mouth to keep out slow birds and low-flying Cessnas, flared wheel arches over the 185-15 tyres with the four yellow disc calipers peeping through nice wire wheels made by some firm called "Leftnearsi Deundo". There is what I understand they call a nerf bar up front — but that sounds like an invention of the speedway grubs — two chrome pipes out back and a flip-lock filler on the offside rear guard.

The dished wood wheel has pukka brass rivets in the rim, the fat black buckets are a little bit too vertical in the squab, and the teensy, itty-bitty shifter with lift-up reverse bar is comfortingly the same as in the TVR-Griffith. There is one forlorn, chromed Godelpusall bar in front of the passenger and an ashtray on the centre tunnel which is highly unlikely ever to be used because the velocity of the slipstream would immediately pluck out any ash and fling it sarcastically into the driver's eyes.

None of this safety crap in the Cobra; your daring pilot sits high, peering over the top of a short screen through streaming eyes, rubbing an elbow on the TC door lock.

In front of him the instruments are scattered

The cockpit, stark, high in the old tradition, well carpeted and assuring seats. Steering wheel is alloy with wood rim and real rivets. Nice touch.

The rear "nerf" bar (as the Detroitonians refer) follows the contours of the tail section — a sight most only see, anyway. AC badge is subtly small.

casually over the dash — speedometer (180 mph), tachometer (8000 rpm) and clock (five past 12). In the test car (a phrase which means We're Not Sure If It Happens In All Cars; and there I go baring trade secrets again) there was a hole where the cigarette lighter should have been. I dare say it could be found in Geoghegan's bottom drawer; the Parramatta Road gentry are notoriously good at lifting shops. Another feature (sic) on the dash is one of those furshlugginer ignition locks which spins around merrily with the key. They had one on the TVR, and also on the first Cortinas, and also on the . . . only the Pongoes could build 'em like that.

I drove this Handy Tool circumspectly for most of the time, but occasionally the blood roared red behind my eyes and I gave it the big razzamoola. I saw 6000 rpm in the indirects once or twice, solid valve lifters meshing clearly with the bellow of the dual pipes, and despite a limited slip rear end it became feasible to poke the car out sideways and hold it there on power leaving thick black marks out of every corner. On playing Don Garlits and Swamp Rat away from a standstill, the hardest thing was not keeping the car straight but moving the hand fast enough to shift before each gear took the tacho needle on the second trip around.

Ah, everybody should drive one of these cars at least once. I got out of the Detroit TC into a Valiant V8 and gave it the big razzamoola but it just wafted palely on. You may drive an E-Type and feel all boy-wonder, or punt a Maserati and be very Farina, but only in something like this can you put your foot in it and things happen.

It doesn't have all that much suspension; in fact, the massive round girders which support the whole deal probably act as giant torsion bars. It rides like a lump of concrete, leaping high into the air over cigarette butts but smiting concrete joining strips without flinching. It is a raw, red heart-in-the-mouth device that makes boys out of men. How the 7-litre version goes I just can't think. Neither, on reflection, do I really want to know. — *Romsey Quints.*

And this is where all that massive torque goes to . . . a set of 185 x 15 Dunlop SP41, mounted on triple-laced chrome rims. Just as a sporty touch.

The bear has grace, but only when it sits still on fat hind rubber. Detroit makes no pretences — this bear's a real, bright red grizzly.

sweet dreams

AC (289) Cobra

By Jefferson Howard Jr.

Left:
"... great bulges over its slicks and a big, wide mouth ... silver blue paint ... glitter ... wild".

EVERY once in a while a dream comes true. I mean, *someone* has to win the pools each week. Then there was this car, see, kinda mean, you know, with great bulges over its slicks and a big, wide mouth. It seemed to press down on the road at each corner. Under the bonnet a great American V8 with a fluttery throb on idle and a great streak down the road when you opened it up. It's the AC Cobra, a hybrid of Ford Mustang power train and British chassis, the last *real* sports car; a man's machine.

I nearly missed it. They gave up production last year, but down on the Thames beyond Kingston there was one left in captivity. It has silver-blue paint finish which nearly jumps out at you with glitter. A weekend at the wheel was like a date with Brigitte Bardot. I mean wild, man, as wild as you can get.

Push that stump of a gear stick into bottom and tread on the loud. Hold tight 'cos you've no g-suit and this is Cape Kennedy. It really flies like a missile off the pad. Ever heard of acceleration *accelerating?* Well she takes off, yes, then she digs in and she flings herself up the road like the elastic just broke and before you shift into second there's 70 on the clock. Trouble is you spend all your time slowing behind the traffic, or so it seems. No squat, no dive, no roll; this one goes flat, every inch of rubber doing work.

In a line you pick your space ahead, judge the gap, and with instant spurt there you are up front. Quick snick into top, makes no difference if you miss a gear or two in between. Oh yes, she's tractable. From top to bottom, it's a diagonal shift with another 3,000 still to go. That's power, *power* man, like too much, and yet it's not.

Ever wondered about those three dots your side of the solid white line? They're for Cobra power, aerosol urge, and this brute can do it. Cars your way stop in their tracks, relative; cars the other way hang back.

Up back you leave them in a kind of thunder. Like a swarm of ton-up bikes on a bender. Maybe the windows don't rattle, but they oughta. It starts as a waffle, then in comes the crackle and from then on it's power waves,

Below:
"... stump of a gear lever ... metal pedals ... do-it-yourself roof kit."

sweet dreams

kapow! On a tight rein there's a tinkly threshing of machinery, poised. Supertaps open, bombs away and there she was—gone; with just the echo where she used to be.

Wheelspin? Nope. Just the yelp on a slow turn from the grip of the limited slip, but there's too much rubber for it to burn. These slicks really stick.

Can't all be good. You're right, cooky. You arrive everywhere hot-foot with those metal pedals glowing cherry. Comes on to rain, and there's that Chinese puzzle of a do-it-yourself roof kit. Like the man said, second only to re-folding a map. Drips on your knees, steam all round, and hot, hot, hot. But who cares? This is a go-car. It's for driving. Whenever you feel tensed and confined, just blow, man, up the road and back in the Cobra and you can't help but feel better. Not on the National Health yet, but a private patient with funds . . .

Unobtainable and unattainable, out of reach like the stars. I gave it back to the man down-town. It's his prize and he won't part with it again, unless he goes crazy or something. Anyway, it's worth three grand, and we don't have that kind of money.

It was nice while it lasted and maybe we're richer for the experience. Hell no, we just want some more, and without it . . . deprivation. Not just a motoring experience of a lifetime, but an experience of a lifetime no matter. There I was, see, with 85 on the clock when I shifted into second and took off. You should have seen the look on this guy's face, but who cares? □

Right:
"Up back you leave them . . . she digs in . . . flings herself up the road."

Right:
". . . another 3,000 still to go . . . into second . . . 70 on the clock."

Below:
"It seemed to press down on the road at each corner."

1962 Shelby Cobra 260

by John Christy

It was a typical Carroll Shelby exercise.

Ol' Shel had this advertisement running in *Sports Car Graphic* magazine pushing his new high-performance and competition driver's school. To keep from being bothered by the 12-year-olds and other impecunious or merely curious types, the ad required that one send a dollar for further information. Ol' Shel was asking you to pay a buck—a solid, 1962 Eisenhower dollar, mind—for the privilege of having a sales pitch laid on you by mail.

Judging by the response he got, there must have been a scad of people in the early Sixties who had an uncontrollable itch to go motor racing. The return address to which you were asked to send the buck was a Postel, a commercial mail drop, at the corner of Hollywood Boulevard and Bronson Avenue about a block from where the Petersen Publishing Co. —and *Sports Car Graphic*—offices were at the time.

About once a week Carroll would make his appearance at the drop and collect the mail. Then he'd go out to his car, open the envelopes, extract the dollar bills and toss the envelopes into a pouch for his secretary to go through and answer with the school sales pitch and application. With the resulting bankroll in this Texan's pocket, Shel would then saunter into the magazine office and, often as not, invite someone, usually this deponent, to lunch—said lunch usually consisting of two, three or four martinis and a steak sandwich. If nothing else it led to voluble conversation.

On one such Wednesday two pieces of intelligence had crossed the desk. To the normal journalist they were totally unrelated. To wit: the Bristol Aircraft Company Ltd. was no longer going to produce the BMW 327 derived engine which it put in its own sedans and sold to the AC Car Company for its roadsters. The second piece of intelligence was to the effect that the Ford Motor Company was going to make the 289-cubic-inch small-block engine universal, dropping the smaller 260-cubic-inch version. Two occurrences roughly one ocean and half a continent apart. At the time, they couldn't know just how close they were.

Along came Shelby, flipping through a roll of bills thick enough to gag a wolverine. It was lunch-on-Shelby's-prospects-and-students time. Good! Perhaps it was just that time of the week when a three-martini lunch was especially appealing. One of the unfailing aspects of these lunches was that between the first and second martini Shelby would begin a conversational segment that began with: "One day ah'm gonna build the fastest production car in the world—*ratt heah* in California." This day, this lunch, was no different. The second martini arrived and the catechism began: "One day..." Like a proper guest I let him finish the first part which ended along about the time we got down to the olive. For some reason—maybe to keep the conversation alive—I then laid the bit about Bristol engines and AC Cars on him, adding that I thought AC would be turning to Ford Zodiac sixes since a conversionist by the name of Ken Rudd in England had been doing just that with AC Ace chassis.

Shelby waved a hand at the waitress for another round. "Keep talkin' boy," he said. As the lady-in-waiting took away the empties and put down the fresh ones, I dropped the second piece of information into the hopper with the added remark that there ought to be a lot of nicely seasoned 260 Fairlane blocks laying in the foundry yard. It was a two-way teaser —the 260 and 289 blocks are the same and I knew that (1) Shelby's Detroit connections were mainly Chevrolet and (2) he was as impecunious as was I, living as he did off the proceeds of an as-yet tiny Goodyear racing tire store.

By the time I had unburdened myself of that particular bit of conversation the steak sandwiches arrived and we, or at least I, devoted the next few minutes to putting a blotter on the gin. Conversation over coffee drifted off to such subjects as the health of the tire business, who did what to the Shelby school rent-a-racer and who was doing whom that week in the SCCA. Afterwards I went back to the office and tried to salvage an afternoon's work out of the gin-soaked wreckage, forgetting about the luncheon talk completely.

The next day I had some reason to call Shelby but when the call went through, a blonde young lady on the other end informed me: "Mr. Shelby has flown to Detroit."

"Oh? And when will he be back?"

"Oh dear, I'm afraid I can't answer that because he's flying from there to England—some place called Thames-Ditton, I think."

Snap! Click! POP! I could practically feel the circuit-breakers in my skull let go. That lousy, rotten Texas sum*bitch*! I'd gone back to the office and sobered up and he'd obviously gone to one of his rich friends and promoted a trip to take advantage of the information I had laid on him for the price of three martinis and a lousy steak sandwich!

I didn't see or hear from Shelby for more than a month. Matter of fact, after that first bit of pique we forgot about what he was up to. After all, that number had been tried before with little or no success.

Then one day in March a call came in.

"Hey," said the familiar voice, "ah got somethang to show you—c'mon out!"

"Out" meant Dean Moon's speed equipment and hubcap emporium where Shelby rented stall and office space for his Goodyear racing tires and himself. I arrived within the hour and saw the activity in the last and next-to-last stalls. In the last stall was the dynamometer on which was bolted a small-block Ford V-8 on the left rocker cover of which was painted the legend XHP-260-2. But it was in the next-to-last stall that the real interest sat on four tautly laced 90-spoke Dunlop wire wheels.

At first glance it was an unpainted AC Bristol car with a longer nose. However, under the hood was the mate to the engine on the dynamometer, XHP-260-1, and squatting in the cockpit was a busy man bolting the shortest driveshaft in captivity between the Borg-Warner four-speed gearbox and a beefy Salisbury rear end center section. The net visual effect was to transform what had heretofore been very desirable property into a device that had grown *hair* in large bundles. The thing hadn't even been fired up and yet it seemed to sit there growling, gnashing teeth and swishing its tail. Not even Shelby knew what he had by the tail at that moment; about all anyone could tell at that point was that whatever it was, it was *mean*.

What we were looking at was by no means new in 1962. The basic design had been bought by AC Cars from one John Tojeiro nearly 10 years before. It had since been seen with the old AC single-cam six and then latterly with the potent Bristol. In this latter form in the late 1950s, it had been an E-Production SCCA class racer to contend with, virtually owning its class and the one above for five years from 1956 to '61. At $6000 it was an expensive device as well, which slowed its sales more than somewhat, considering the simplicity of the machine.

Derived from a one-off racer, the chassis consisted of a pair of large longitudinal frame tubes with one cross member and two large bridge structures at each end. These bridge structures carried A-arms at the bottom of each end and on top also at each end were semi-elliptic transverse springs that also acted as upper control arms for the suspension uprights. The rest of the frame was small diameter tubing whose primary function was to carry the very light aluminum body panels. As indicated, the suspension was fully independent front and rear. On this particular version there were disc brakes front and rear with the rear discs and calipers carried inboard next to the rear-end center section. Later ones were to have the rear discs located outboard because the heat from the brakes augmented the heat in the center section and vice versa, with all sorts of unhappy results.

The cockpit was sparsely covered in fabric and leather and furnished with a pair of superbly comfortable and form-fitting leather-covered seats. On the leather-trimmed dash there was a positive welter of instruments. Though laid in with no particular pattern, there

RETROSPECT

1962 Shelby Cobra 260 (Number 1)
Owned by Carroll Shelby
Photographed for *Motor Trend* by PPC Photographic

were enough of them to satisfy all but the most crass instrument freak. Almost vertical in front of the driver was a medium-sized wood-rimmed wheel and on each side of the steering shaft were a pair of antiquated-looking pedals that appeared to have come from a motoring museum.

Running down the center of the cockpit was a high central tunnel from which protruded, just below the right rim of the steering wheel, one of the shortest gearshift levers in the memory of racing man. The whole thing was, in short, derived from a racing car and looked it. The major changes wrought at Shelby's behest were longer springs and A-arms to give a wider track, bigger brakes, a heavier Salisbury center section and drive train and heavier locating rods to take the added torque. The weight of the A.C. Bristol was about 1900 pounds and the new Shelby version (no one had come with the name *Cobra* at that point) weighed in at 2100 pounds. The price per pound was down, though, because the list price was $5995 plus tax and license.

To get back to the original chronology of events, I had to leave that evening before they got the beast together but I was told to come back the next day and pick it up "if it's running right and still in one piece." I called ahead the next day and was told that everything was fine. Seems that Shel and Dean Moon had gone out drive-in haunting and Corvette-hunting the night before but found no game so the first night's exposure index was low. I dashed out and took the car off Shelby's hands *before* he got any other enthusiastic bright ideas *before* we could extract some driving impressions and photos, some of which are reprinted herein.

By any standards—today's, those of the muscle-car era of the late '60s, or those of 1962, that car was absolute dynamite. It didn't have a name as yet but it had *style*. Totally innocent of paint with the name "Shelby" scrawled in paint on the nose and tail, looking about as raw as a car can look, it still had *style*, something that made heads turn. Aside from that, it would leave any car on the road standing in its own dents. It had a unique quality that I've seen in few other cars: It got underway in an absolutely blinding rush without any untoward wheelspin or rubber burning, or sideways cop-attracting nonsense. You poked the pedal and dropped the hammer and the result was instant motion foreward. The whole thing was a revelation, and while it may have seemed a bit raw and crude to those who were into sophisticated Grand Touring machinery, there was no doubt that this was truly, in every sense of the word, a sports car. It was, in fact, a rolling, moving definition of that much abused and confused term—if you ever need to know what the term means, drive a Cobra.

Shortly after that first session—in which we found that 260 horses from Ford would propel 2100 pounds of AC roadster at a rate of 150 miles an hour—the car was painted a brilliant pearlescent yellow by Dick Jeffries. In this guise it went to the New York Automobile Show under the sponsorship of the Ford Motor Company where it absolutely stole the show from the rest of the FoMoCo exhibits and those of other companies as well.

Before that spring and summer had ended, two more were worked on with the wars of competition in mind. They were in and out of the shop, back and forth to Riverside and Willow Springs. In the fall, one made its public debut at Riverside in the hands of Billy Kraus. Thanks to a badly designed chamfer in a rear hub, the run was short lived but very demonstrative. While it was running, Billy kept it ahead of the hottest Corvettes on the West Coast. Later that fall, the two cars were taken to Nassau where they demonstrated that they could stay ahead of even the likes of the then-redoubtable Ferrari GTO. One Cobra bid fair to take the Governor's Trophy away from Ferrari until a miscalculation on how much gas a Cobra needed to run an hour forced a halt to the proceeding.

Cobra Number One, however, was not to join these battles. While others went on to take everything in sight that the SCCA road racers could field, it went back on the street. The pearlescent yellow couldn't take the California sun and turned all leprous looking, so it was exchanged for Guardsman blue which had become the Shelby team color that it remains to this day.

In 1963, the Cobra took home the SCCA B-Production championship. In 1964 they made a run at the Manufacturers' Championship, missed, but took the class in SCCA again. Shelby made a public promise to the effect that "...next year, Ferrari's ass is mine."

On July 4, 1965, Bob Bondurant and Jo Schlesser drove a Cobra 289 to win the Grand Touring competition at Reims, France and thereby cinched the World Championship of Makes for an American manufacturer for the first time and the Cobra rolled into the world's history book. Ol' Shel did indeed build the fastest production car in the world. It's still right up there. ■

1.
If you can see that sharply, the speedometer originally in Cobra I was a Mercedes unit; the car was delivered without one. Cockpit was sparse but snug; seats were superb.
2.
Engine in Cobra I was the first high performance small-block Ford. It put out 260 honest horsepower at 5800 rpm, enough to propel the car to a speed of 150 mph at Riverside.
3.
In spite of the fact that the horsepower had been doubled over that of the AC Bristol the Cobra cornered better than ever with none of the lean of the early AC.
4.
Braking, for the early 1960's was awesomely strong and fade free thanks to disc brakes on all four corners. The lack of upper locating link didn't seem to hurt.

RETROSPECT

1964 Shelby-American Cobra Daytona

Gentlemen, start your legends

by Len Frank

PHOTOGRAPHY BY BOB D'OLIVO

We live in parlous times. The Rough Beast slouches through all the world causing us to Doubt and Despair. Captain Marvel is forever bound in litigation. Pennies are made with zinc, quarters are made with copper; the coinage is debased. Volkswagens are built in Pennsylvania, Hondas in Ohio, Kawasakis in Nebraska, Datsuns in Tennessee; our borders have been breached. Red, white, and blue are bands on the tricolor and they soon will own AMC. That's *American* Motors. What's left?

Well, we still have heroes; we still have legends. And if those are still terribly important to you—if you want, like Rick Nagel, to remain a believer—turn quickly to the large color photo, look at that brutal winner, make plans to visit the museum, then pass on.

I have not met Mr. Nagel, but he seems a nice fellow (on the telephone). He bought CSX 2602, the coupe shown here, from Mike Shoen who owns another coupe, *and* an FIA roadster, *and* the original car raced by Billy Krause at Riverside (or at least he still had these when last I checked). Mr. Nagel, too, has a considerable collection in Dallas, including memorabilia and vehicles peripheral to the Shelby operation. He hopes to open a museum—a memorial to that hero, that legend, Carroll Shelby.

On one of those chatty afternoon TV talk shows—with the content and impact of a supermarket kitchen-and-closet magazine—ol' Shel was a guest, come to talk about chili. It was conceivable that he might have been there to talk about Pit Stop Deodorant, but that seems to have faded away of late. Shel, always gracious with the ladies, was opining about *real* chili when the hostess, a gusher, an interrupter, asked him if he hadn't been a race driver "or something—didn't you set the world's speed record or something?" Shelby said the appropriate things about Le Mans, about the old days, then steered smoothly back to the chili. Maybe the museum will have a package of Carroll Shelby's Original Texas Brand Chili Preparation (World's best, makes 2 qt) net weight 4 oz.

Unlike the gushing lady interlocutor and the modest Mr. Shelby, most of those who care are steeped in the lore of the Cobra. They know that it was the fastest, meanest, baddest car in the world then, now, or at any time up to the advent of the nuclear auto or the nuclear holocaust.

The Shelby-American Cobra had improbable ancestors, but then most Americans do. Way back, it counts the cute little Fiat 500A, the Topolino of the '30s. The Topolino is one of those cars that seemed to have been designed just to inspire specials builders/hot rodders. In this case, a couple of Fiats gave their front ends so that the tiny Cooper F3 car might have 4-wheel independent suspension.

A bit later, Cooper built a front-engined car, one of its few, with a 2-liter Bristol six. The Bristol was, of course, a development of the prewar BMW.

Retrospect: 1964 Shelby-American Cobra Daytona

Photographed for *Motor Trend* by Bob D'Olivo   Owned by Rick Nagel

Cooper had an employee, John Tojeiro (of Portuguese ancestry) who struck out on his own and used the Cooper-Bristol formula to build an MG-powered sports car that was eventually rebodied to resemble a touring-bodied Ferrari, the type nicknamed *Barchetta.*

This became the prototype of the A.C. Ace, later, with an engine change, the A.C.-Bristol (they already knew the combination would work), and, when the Bristol engine went out of production in 1961, the A.C.-Ford (the Falcon-like English Ford Zephyr 2.6-liter six). Be patient—only a moment longer.

In a display of the skills that Shelby had only been polishing as an international race driver/*bon vivant*/down-home striped overalls genuine Texan chicken farmer, he essentially told A.C. that he could provide engines, and Ford he could provide chassis. If we had some ham we could have ham and eggs, if we had some eggs. He was good at it.

About four years and 1010 Cobras later, the production, but not the continuing saga, was finished. That 1010 figure perhaps does not include British oddities or the occasional counterfeit, but it does include 260 and 289 small blocks (Raviolo-designed thinwall V-8, originally 221 cu in. from the Fairlane and Falcon Sprint) powering street roadsters, Le Mans roadsters, Le Mans replicas, FIA roadsters, USRRC roadsters, 427 big-block cars, and, of course, a half dozen Daytona coupes.

More hyperbole, more misinformation, more distortion, exaggeration, more sheer nonsense, has been written about this slim production than about any other similar number. And maybe this too is hyperbole. What's another pebble of nonsense in a landslide?

The second Cobra built was raced at Riverside in October 1962. Billy Krause, a local driver with international talent, was at the wheel. Krause led until a rear hub broke. The Cobra really was rolling proof that "racing improves the breed." By the time that last 289 was built (the 427 was a wholly different car), the frame, hubs, stub axles, steering, cooling, body work, brakes, and wheels reflected racing development.

A Duesenberg driven by Jimmy Murphy won the Grand Prix held at Le Mans in 1921. Despite valiant attempts by Miller, Stutz, Chrysler, Cunningham, and finally Scarab, an American car did not win a major European road race until the Cobras, 40-odd years after Murphy. Legends.

Despite dominating Sports Car Club of America production road racing and doing well enough as USRRC (predecessor to Group 7 Can-Am racing) cars, the roadsters had not had significant success in Europe. Their parallelogram front and rear suspensions required brutal roll stiffness to keep the ever-widening tires flat on the ground. Top speeds were limited by basically 1949 (*Barchetta,* remember) body shapes made worse by the FIA windshield requirements.

Pete Brock had really wanted that first ride in the Cobra. He was like so many of us drawn to Southern California because it was all here then. Brock, trained at Art Center School (now Art Center College of Design), had worked under Larry Shinoda on the Stingray at GM.

The Cobra was rolling proof that racing improves the breed

Dissatisfied, he packed up his Le Mans Cooper 1100 and migrated to Riverside "because Gurney and Skip Hudson had started there" and bought a little house across the road from the outside of turn nine. Max Balchowsky, a man for whom Brock still shows great respect, gratitude, and admiration, was teaching Brock the rudiments of tending the Lotus 11 that replaced the Cooper. Balchowsky effected an introduction to Shelby—this was sometime in 1961—who was visiting Max looking for the Way of Wisdom that would lead to his personal grail, the Cobra. Max was the designer, constructor, and sometime driver of the Old Yellers, the quintessential American chalk-on-garage-floor specials.

Shelby was at something of a loose end after his retirement from driving and the scheme to produce a sports car was just one straw to grasp at. The driving school was another one. Shelby and driver Paul O'Shea had gotten together to open a school of high-performance driving, the country's first. Their partnership, says Brock, "lasted about two days." And Brock, hanging around the raceway, remembers Shelby walking up to him and saying, "you're the driving school."

Ken Miles, another Southern California (by way of England) legend, came to Riverside to do some development on that No. 1 Cobra. Brock says that Miles taught him "everything I knew about driving in three days." Miles deserves a story—maybe a book—all his own. But for now know that he was the father of the 427 Cobra, the Sunbeam Tiger, sometime driver of the late Chuck Porter's SLS (see Abarth 207A Retrospect, Nov. '82), the original chassis designer of the Reventlow Scarab, America's fastest Porsche driver, an absolutely fierce competitor from Le Mans to Sebring to Riverside in anything from an MG-TD or Sunbeam Alpine to the Ford J-car. It was

in testing the latter in late August 1966 that he left the end of the straightaway at Riverside at very high speed and was killed.

Brock really wanted to become a professional driver and was crushed when Shelby chose Krause to drive that first race. With all the time on the track in a Cobra, Pete could lap faster than Krause, at least on an empty track. He realizes now that Krause would certainly be faster in traffic. Hub failure made it moot.

Aside from actually racing, Brock's only other ambition then was to design a car. Shelby kept him running the driving school (now with John Timanus) and doing posters and other graphic designs.

Shelby, by this time, had taken over the remnants of Lance Reventlow's RAI in Venice, lock, stock, Phil Remington, and junk box.

The early '60s was an interesting period in motorsports. It was the last stand of the front-engined car, a period of dominance for Ferrari as Talbot, Mercedes, Jaguar, Alfa, and Maserati all pulled out of racing. New entrants like Lotus and BRM were involved only with Formula cars—Ferrari ran Formula One, Formula 2, Prototype, Sports, GT. Porsche had not yet emerged as a competitor for overall wins on most courses.

Ford saw some opportunity to enhance its image through racing and made an offer to buy Ferrari. Not too surprisingly, it fell through. It then began a racing blitz of its own, absolutely unmatched ever. Within a few years it had become a force at Indianapolis and Le Mans, on drag strips, in NASCAR, the East African Safari Rally, and Monte Carlo Rally. It paid for—or developed—the SOHC 427, side-oiler pushrod 427, 4-cam alloy Indy engine, Ford-Lotus (Cortina) Twin Cam, Formula Cosworth DFV—the list is almost too long. What it lacked in experience Ford also built or bought.

When Shelby, Brock, Miles, Remington, and company really got going, Ford was heavily involved with Holman-Moody-Stroppe and the development of the Lola-based GT-40. But despite the truckloads of cash Ford was throwing at its GT-40 effort, it was the comparatively small Shelby operation that was winning.

Brock approached Shelby and enthused about the potential of a more streamlined Cobra to actually contest the World Championship of Makes (for GT cars). Brock remembers Shelby being quite offhanded about it all, but he approved the project without much discussion. Brock also remembers Remington being pretty negative about the project. Miles, however, jumped into it.

Miles began a series of chassis modifications designed to lower the driving position and center of gravity, and to stiffen the A.C.-based frame. There was no time for a clay model, no time for wind tunnel testing. Brock just began to draw, and, with suggestions from Miles, had working drawings in a remarkably short time.

Remington doesn't remember being negative—just busy. By this time Remington had had European racing exposure both with RAI and Shelby. He had gained respect for Ferrari, especially the engines, though he remembers that the pieces they bought instead of making themselves were "junk." He also had gotten a look at the GTO and GTB, and knew that Ferrari had a mid-engined GT car cooking (the LM series—not homologated). Shelby might make noises about "gitting Ferrari's ass," but Remington had some clearer idea of what it would take. And he was pretty sure that a 1952 chassis with an overstressed production car engine, with or without a coupe body with questionable aerodynamics, was not the answer.

In motorsport, the early '60s was the last stand of the front-engine car

Miles talked Brock into less windshield rake than he wanted to use, on the premise that, in a 12- or 24-hour race, it was far more important for the driver to be able to see than for the car to have a few mph more top speed.

Brock tried to get some of the shape back by moving the high point of the roof aft of the normal windshield header high point. It gave the car a slightly strange, unfamiliar, humped, sinister look. Actual aluminum work was done by California Metal Shaping, finished in the Shelby shops.

An adjustable loop airfoil, a Brock trademark, was not built because, Remington says, of time limitations. The loop airfoil was not built, Brock says, because Remington thought it was more foolishness. The primary constructor on the car was a New Zealander, John Olson. Brock credits his dedication with the completion of the first coupe. It was Olson who worked out many of the small details that make the difference between finishing or not—often the crew chief on a car is more important than the engineer or driver.

At one point during the construction, Shelby brought Benny Howard by to look at Brock's pet. Howard was the designer of Pete, and Ike, Mr. Mulligan and the DGA ("Damn Good Airplane") series, Bendix and Thompson Trophy racers, and "the Lear jets of their day," says Brock. Any-

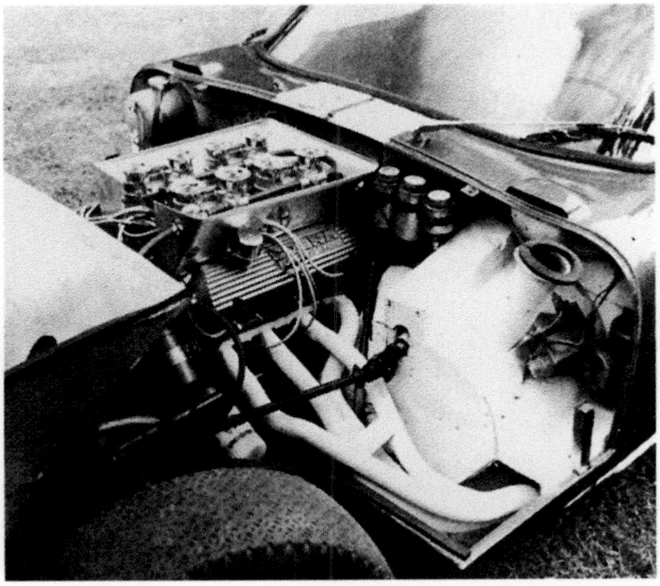

way, Brock remembers, Howard took a look and said it was all wrong, that it would take about 450 hp to go 150 mph. They had 375/385 hp and they needed 170 mph. Shelby and Howard went to lunch.

Remington says that Shelby was really a "good news guy," that he would explode if something went wrong. Remington, now fabricating cars for Dan Gurney's AAR, remembers that "Shelby had very little input. He relied on our judgment. He just wanted results."

Shelby came back from lunch, asked Brock what about it. "Trust me," Brock said. Shelby did.

A few minor problems did develop. The rear quarter windows were to act as rear brake air scoops. During the limited testing time, Brock found that the rear wheels acted like air pumps and pushed air *out* through the *in* scoops. Plexiglass deflectors by the A-pillars fixed that. The car was several hundred rpm faster down Riverside's straight than a roadster with a better engine. Obviously it worked. Later, to aid stability, a ducktail spoiler was added. Brock still thinks it would have been faster with the loop spoiler.

Okay, Pete Brock now heads a modestly successful hang glider firm (UP Industries, Temecula, California). Remington, as we said, works for Gurney. Miles is dead. What about Carroll Shelby, the ex-chicken farmer, ex-race driver, ex-ham-and-eggs optimist turned chili mogul. Well, he still heads Shelby-American. After Ford bought out the Cobra operation, Shelby signed a development contract with Toyota to run the 2000GTs. Brock, off to run BRE, remembers the period as the heyday of SCCA production racing with Brock running a Datsun team, Richie Ginther Porsches, Kas Kastner Triumphs, Joe Huffaker MGBs. Brock had run a team of Hino Contessa sedans before the Datsuns. Hino was absorbed by Toyota; and with that and Shelby's association with Toyota Brock made a deal with Datsun, he says, "just to get even."

Brock ran the Datsun 2000 roadsters, the 240Zs, and the Trans-Am 510 sedans.

It had improbable ancestors, but then most Americans do

He says that in the process, he learned why Shelby made some of his decisions. Brock emphasizes that at no time did he have to stay with Shelby, that he generally loved what he was doing, and a good thing too, because he made very little money doing it.

He recalls Shel's early PR activities when he was a "good ol' boy racer" keeping up the hooraw with the small town sportswriters across Texas. Brock says Shelby never forgot those guys, always kept them apprised of his activities, sometimes real, sometimes, well, a bit exaggerated. It kept the writers happy, it kept Shel supplied with press clippings. Not that he was all puff. He drove some rotten "hawgs" and made it look like fun.

Dan Gurney couldn't remember for a bit what it was like to drive a Daytona coupe or even whether he had *ever* driven one (you did, Dan—you won at Le Mans in 1964 with Bob Bondurant—there were other times). But he eventually recalled that the roadsters "were more fun to drive in a way—you could throw them around." He, like other Grand Prix drivers, was not in love with distance racing where the car had to be spared, but he says it "was a heady period in a way." He liked the patriotic aspect of driving the American cars.

Brock says that he is still proud of the Daytona, but then he's proud of every car he ever designed.

About Shelby, he concludes, "He's just Shelby." MT

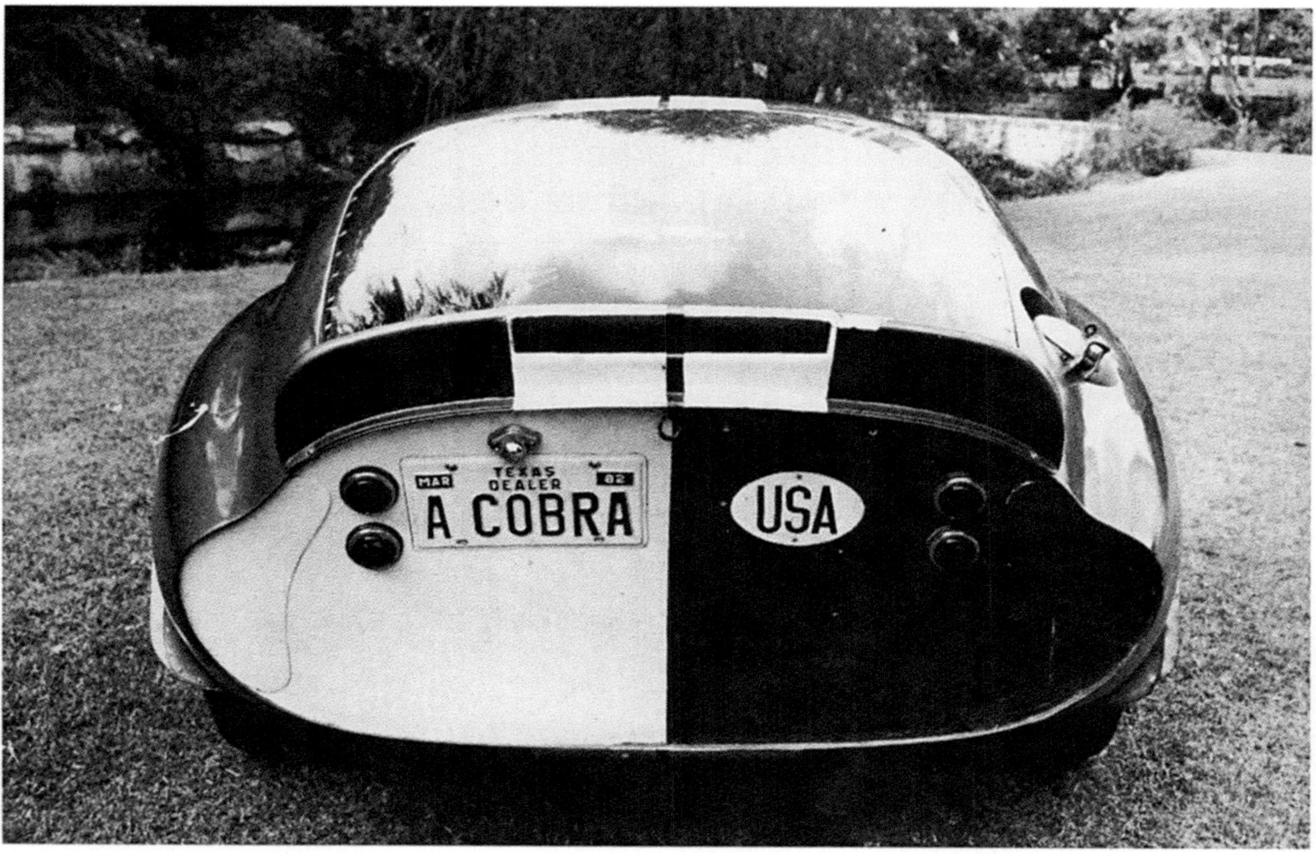

A particularly Brief Test

AC Cobra 427+

This is necessarily a Brief Test. It is actually the Briefest Test *Autocar* has ever conducted, for various reasons, connected with vague quantities like 427 cubic inches, allegedly rather more than 500 bhp and a little more than 1 ton, 0 to 60 in 4.8, and a tyre that went bang in the day.

IT all began when a phone call came from Hexagon of Highgate Ltd., North Hill, London N.6. They buy and sell cars, mostly sports and high performance cars. They had imported back from the United States a pair of 7-litre AC Cobras, the last of which were built in 1968. Both were in exceptionally good condition, they said, a red one with an unmodified engine and a silver one with such items as something called "tunnel port heads", a funny camshaft, and fat wheels and tyres. Would we care to try the latter?

We were told that the silver car had been bought by, of all people, a man employed by General Motors. He presented it to his son, who decided that in standard form it had too little power. The modifications were carried out, but for some reason or other he didn't drive it much. Stored in some place, vandals had broken in and done their best to destroy it, but fortunately not very well. Fully repaired, and with only 2,000-odd miles on the clock, it was then sold to a dealer, from whom Hexagon bought it.

Everything is simple on the Cobra. One cannot help being reminded of the humble Austin-Healey Sprite as it first appeared, surprising as that may seem. Rather more instruments of course, but set in a similarly plain panel, very AC with its high hooped top. Rather more litres too. As performance figures would be best taken with the hood up (for aerodynamic reasons) Hexagon erected the canvas for us, and that proved to be the same old affair of loose hood sticks and Lift-the-Dots; it proved to be perfectly adequately strong for its job though. None of that wind-up window sloppiness either, but a pair of flappy detachable Perspex sidescreens, plus hinged draught deflectors at each end of the screen like the things you find on early Vintage-period touring cars. Push-pull and unashamed toggle switches, unperplexed by the safety fuss, stick out of the panel, some of them confident that you'll find out what they're for when you switch them on, rather than bother with labels or symbols. It is, as Hexagon agreed, a throwback, a post-war Vintage car, in most of the best possible ways.

From a standing start

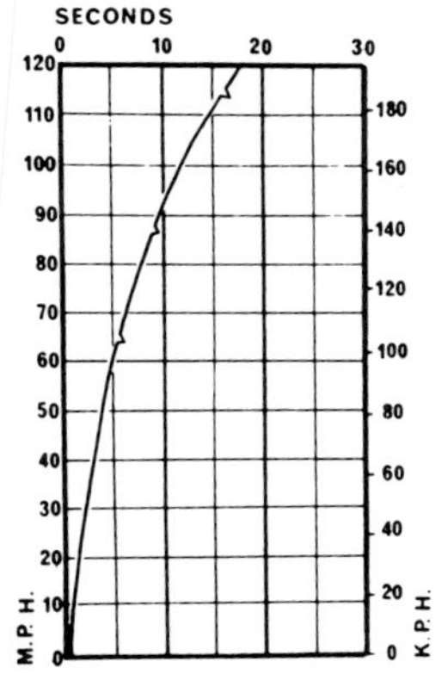

MPH	SEC
0-30	**2.2**
0-40	**3.0**
0-50	**3.9**
0-60	**4.8**
0-70	**6.6**
0-80	**8.0**
0-90	**9.9**
0-100	**11.7**
0-110	**14.5**
0-120	**17.9**

standing ¼ mile 12.9sec 105mph
standing km 24.0sec — (see text)

Top: The back tyres spun nearly all the way to 60 mph in 1st gear on the straight at MIRA

Below: Another standing start accompanied by a great deal of oil and tyre smoke

It is made for a slightly shorter man than we are, but is in the way of most sports cars very comfortable nevertheless. You sit beside the big fat transmission tunnel on one side and the delicate almost shell-like aluminium-panelled tubular framed door on the other, but unlike its sports-car contemporaries the seat is not very low so that you seem to sit more on than in the car. The gross bulge of the rear wheel arch puts road filth on your clothes as you get out. The front wheel arch bulge seen from the driving seat beyond the swoop of the wing itself is perhaps unintentionally but suitably reminiscent of the hood of a real cobra. It must have one of the most impressive views from inside to be seen in any car except perhaps the E-type Jaguar — all bonnet and wing, each only just hiding the size of what's underneath. Mid-engined cars will never have the appeal of such beasts as this in the same way. Underneath it all is the simplest of chassis, two huge tubes, cross braced with other similar tubes, ladder-fashion down the middle of the car, with built-up tubular structures at each end for the nowadays old-fashioned 2½-litre-Cooper-style all-double wishbone coil-spring independent suspension.

Driving it

Switch on, and the click is heard of the Bendix fuel pump. Turn the key, prod the throttle two or three times to wet the carburettors' eight throats and with a tremendous roar the engine bursts into thunderous life. In fact, the first time I did that, what had seemed to be a suspiciously stiff throttle pedal stuck fairly wide open. By experience wary of such cars, I switched off again very quickly, at the same time trying to pull the pedal back with my toe; a throttle spring was changed, and all was well. Even though the engine has a special steel crankshaft and a number of other improvements to enable it to sustain high rpm, the red line was set at 6,400 rpm and it was immediately obvious that the rate at which the free-running unit would

accelerate was astronomical. I was glad I switched off.

Once warm, it was something of a surprise to find that the thing would sort-of tickover. Tickover consisted of loud bumpedy-bumpedy-bumpedy-bump noises, like one of those early stereo demonstration records, very much noises on each side of you, due to the fact that the big twin exhausts ended just aft of the back of the cockpit. The clutch was very heavy, and so were the brakes, not surprisingly as no servo is provided and the non-ventilated Girling discs looked a little small by modern standards. The steering was heavy, kicked back in the usual AC way, and was full of feel.

Driving it very gingerly at first to the beginning of M1 trying to think how I could make myself look and sound inconspicuous — which was rather like bringing the *Queen Elizabeth II* into the Pool of London and then pretending that it wasn't there — was a bit difficult. The Cobra is conspicuous. Putting up the hood helped slightly to hide the large chromium plated roll-over bar, which the black canvas just covered like a piece of suggestive feminine underwear. The noise from the exhausts plus, at low town rpm, the tap from the solid valve-gear (as opposed to pansy hydraulic lifters) was prodigious. It would have been far more interesting to drive the car on ordinary roads the whole way, but M1 offered a way clear of any human habitation which was most desirable. The only initial worry was temperature; in chilly English winter weather, the water temperature wouldn't rise higher than 140 deg F and the oil temperature gauge, once on the motorway, dropped from about 110 deg F back to the stop, which is at 90. AC hadn't made the old British exporter's errors of underestimating Californian ambient temperatures, so that now the car was over-cooled.

At MIRA

It was good to get there — an end to a lot of restraint. We fixed on the solid bar used on very fast cars in place of our bumper clamp for the attachment of our fifth wheel, put up the tyre pressures, then headed for the banking to make sure the car was as warm as possible before attempting any standing-start acceleration tests. Then we made for the twin horizontal miles for the interesting bit.

I had already learnt that there was a tendency for the back wheels to spin. Believing this to be wasteful, I intended to use something like 3,000 rpm for the take-off. I had also learnt that the power started to come in at about 2,500 rpm, becoming most marvellously violent from 3,000 onwards all the way to 6,000 odd.

So, with David Thomas on the watch, rev counter reading 3,000, first gear engaged, the clutch was let in. It was initially most irritating — the revs dropped with a monstrous cough, the car hung whilst the driver swore, smelling evil whiffs of clutch lining, then abruptly everything started working. There was the most enormous blast of sound, hard to convey in words, the great blistered nose lifted again, the big tyres spun, the car twisted slightly to the left, the rev counter needle shot round the scale while the driver tried to watch it carefully *and* keep the beast straight and change gear decently as we were propelled up the straight at a rate so consistently powerful that it took our breath away. The gearchange was the only slow thing on the car, having a long travel. Nevertheless, in spite of nearly a second lost at the beginning, David was shouting at me to lift off 18.9sec. after the start and well before the kilometre mark for fear of bending the needle against the stop on our old spare fifth wheel speedometer by exceeding 120 mph. (Unfortunately both our normal 160 mph speedometers were away being serviced at the time.)

The Cobra's Cockpit with a big fat transmission tunnel and push-pull switches on the facia

7-litre of vee-8 power provided the quickest standing start times to 120 mph we have ever measured

Below: The burst rear left tyre, which cut short out test, failed the bodywork causing the paint to blister

Pointing the other way, we tried again. Preferring wasting rubber to fading clutch lining, I set the rev counter pointing at 5,500 rpm, and let the clutch in with a bang. There was no mistake this time. The back tyres spun nearly all the way to 60-odd mph in 1st which was roughly where we changed gear (6,000 rpm) — we paced out the black marks later and found that they went on for a conservative 60 yards. The rest of the run was the same as the first one — the incredible combination of that truly mighty engine's monstrous voice with the most exhilarating yet steady shove forward for the whole run, plus, now that I was more confident of the Cobra's manners when provoked, the sight in the mirror of a great deal of oil and tyre smoke rising behind us over the first bit on which the car had swung even more markedly leftward. There is no point in repeating the figures here except to say that they are the quickest standing start times to 120 mph that *Autocar* has ever measured. Principal points of comparison with the fastest road-going production car we have so far tested, Ferrari's 365 GTB/4 Daytona (*Autocar* 30 September this year) are as follows, with the Ferrari times in brackets: 0 to 60 mph 4.8sec. (5.4sec.), 80 in 8.0 (8.6), 100 in 11.7 (12.6), the $\frac{1}{4}$-mile in 12.9sec. at 105 mph (13.7sec. at, interestingly, 104 mph) and, with Ferrari aerodynamics beginning to tell, 0 to 120 mph in 17.9sec. (Ferrari 18.0sec.).

We were impressed. A really practised AC driver would probably do better, as this one was not anxious to practice any karate gymnastics on the somewhat domestic gearchange with such a lively set of pistons only too disastrously willing to pop out of their cast iron roofs if he missed. They probably wouldn't just pop either, although David Thomas seemed to recall a man with a similar engine who is happy to rev it to 8,500; but then he probably owns the engine, which is a different kettle of gudgeon pins. Remarks like "fantastic", "there ain't no substitute for cubes", "cor" etc were exchanged.

After one has been testing motor cars for a little while, one remembers at the back of one's mind occasions when, after a series of successful acceleration runs, one has said "Just once more, to make sure" and set off on a run that has proved mechanically fatal. Provided one is sure that you have done your reasonable best, it is always better to stop. I recalled this while saying that we ought to do one pair of runs with the electrical feed to the speedometer head disconnected, so that we could get an idea of what the time over the standing kilometre was.

Off we went again, with that glorious snake left, all the sound and the fury — an overworked phrase which truly does apply in this case — and the delicious pleasure of keeping one's foot flat down beyond 120 instead of frustratingly having to lift off. The $\frac{1}{4}$-mile came up in the now usual 13sec. area and we roared on; the kilometre streaked by and I was by now having to worry about the possiblity of over-revving in top. The rev counter pointed to 6,200 rpm which is the equivalent of 143 mph at somewhere around the $\frac{3}{4}$ mile mark and I was just thinking that that was enough when there was a loud report behind my left ear. The left rear tyre had burst, as we both realized straight away, but the car was fortunately so stable that, with carefully applied braking we stopped nicely in time with no undue wobbles on the beginning of the banking at the end of the straight. That is the final reason why this is a somewhat brief test; by the time we'd persuaded the car on to a succession of jacks — ground clearance is limited even with all tyres inflated — and had fitted the wrong-size (standard) spare wheel, there was obviously no chance of finishing the test properly.

We tidied everything up, refitting the car's back bumper. The others went off while I rang the office in order to ask Geoffrey Howard to warn Hexagon that I was going to be late returning the car.

Coming out of the MIRA control tower late on that winter evening, the scene was almost Christmas-peaceful after all the excitement. A near-full moon was well up, silvering everything. Everybody had gone home except for one long lorry banging desolately along the pavé circuit, with the ice-warning lamp on the high-speed circuit blinking blue in the distance. The Cobra sat there, a silver sprawly bulgy shape silvered in the unearthly light, the moon glinting on the high points of its body, an animal sleeping more than a reptile but warm and waiting to move. It seemed almost sacrilegious to wake its rumble again, but the moment of re-starting was one of those odd ones in life, something to savour. What a way to spend Christmas. Cor indeed. □

DON'T MESS WITH JIM McCASHIN'S 427 COBRA UNLESS YOU WANT TO GET . . .

SNAKE BIT

By Gray Baskerville

HOT ROD MAGAZINE ROD TEST

The wide-eyed stares from the two young dudes in their clapped-out convertible said it all. Their lips moved in perfect unison as both mouthed the words, "No bleeping bleep, a four-twenty-seven C-O-B-R-A!" They too, like the gas station attendants in Malibu, the innumerable sporty car drivers along Pacific Coast Highway, and the portly fellow from Pasadena, had been felled by the pearl silver and blue brilliance that remains part and parcel of this fastest of all production roadsters.

Yet an entire generation of car enthusiasts has never even had a glimpse of one of these aluminum-bodied, tube-framed, two-seater, semi-streetable racing cars. The last one came off Carroll Shelby's slow-moving production line more than ten years ago. And even fewer would exist today if it weren't for the happy fact that Shelby had to produce 500+ of these nasty, overpowered roller skates in order to qualify for the '66 Manufacturers Championship. He did and they won . . . just about everything they entered. Back then, $5000 would buy you a tractable street version in a 427/4-speed configuration that was guaranteed to knock the socks off any Vette or GTO around. For another 500 you could order a Shelby Cobra in semi-race form. And if that wasn't enough, an additional half-thou got you the full-race number, complete with rearend oil coolers, close-ratio gearboxes, roll hoop, and a Holman and Moody-modified medium riser that spun out 485 horses and 510 foot-pounds of torque. When you piled all of this into a 2200-pound package, added some wide rubber and seasoned it with a minimal amount of add-ons, you had something that would run in the tens at the drags or 1.35s around Willow Springs.

Jim McCashin was one of those fortunates who had the long green wherewithal to buy one of Carroll's Cobras. In fact, he managed to acquire three, one of which ended up as a turn-one total at Riverside. But not all was lost. The very fact

PHOTOGRAPHY: PAT BROLLIER, PPC PHOTOGRAPHIC

that by 1970, 427 SC Cobras were fast becoming junk or collector's items caused the fledgling attorney to mend his ways. He was able to mine the total of most of its running gear and powertrain while lucking into an almost perfect body and chassis combination. For seven back-burner years, McCashin, engine builders "Cas" Casner and Dave Dralle, plus aluminum-pounder Dave Kent, slowly restored the basket case into one of the finest members of the marque this side of El Segundo.

There should be no doubt that a 427 Cobra is and/or was designed to go racing. It hasn't any provisions for power-assisted options, cruise control, side windows, air conditioning or factory stereo/CB. There is no such thing as factory warranty, five-mile-per-hour bumpers or environmental protection devices. No, a Shelby Cobra is simply a race car with enough legality built into it so you could drive it on the street.

And as with any race car, you don't exactly jump into either the driver's or passenger's seat. The technique is to gently pull down on a leather strap that is attached to the latch, then gingerly slide in, avoiding the black chrome collector along the side of the roadster. Once there, you are ensconced in the minimal bucket seats, surrounded by real leather and looking at one tiny glovebox, a gaggle of toggle switches and a set of partially working Smith gauges. Then you wait for the thunder that you know is coming. The bellow begins immediately and your nose starts to pick up that familiar odor of hot oil fumes mixed with spent exhaust gases. Quivering from oil cooler to bumpers, we ease out of the wicker basket and onto the Santa Monica Freeway with the elan of a moon probe leaving the pad at Cape Canaveral. And once the 11½-inch single-disc clutch has engaged and all three gear changes are com-

SNAKE BIT

There it is, folks, in all its chrome-plated glory. Under the glitter is 427 inches of rompin', stompin' medium-riser Ford. Compression is 12:1, heads have been fully ported, polished and cc'ed, Ford medium-riser intake supports 850-cfm Holley, and everything is sent through an 11½-inch single-disc clutch.

Visibility unlimited, but don't look for full cruise, climate control, power steering, ten-way seats, three-on-the-tree, stereo or CB. Jim's Cobra isn't designed to carry any kind of creature features except those that add to its performance. Simple buckets are covered by black leather, as is the dash. Smith gauges work when they want to, glove box has just enough room to carry one street map, and carpeting barely keeps out the heat.

HOT ROD MAGAZINE'S
ROD TEST SPECIFICATIONS
1966 427 SC Cobra
Jim McCashin, Los Angeles, California

ENGINE:

Engine Type	1966 Medium-Riser Ford
Displacement	427 cubic inches

DRIVELINE:

Transmission	1966 heavy-duty, close-ratio Ford 4-speed
Rearend	1966 Salisbury running 3.77s

CHASSIS:

Frame	Large-diameter steel tube
Wheelbase	90 inches
Front Suspension	A-arms of unequal lengths using large coil springs, hydraulic shocks and sway bar
Rear Suspension	Fully independent Salisbury with limited-slip differential, with coil springs and hydraulic shocks and sway bar
Spindles	1966 Shelby
Tread	56 inches front, 58 inches rear
Brakes	Girling disc brakes, 11⅝-inch front, 10¾-inch rear
Shocks	Koni adjustables front and rear
Springs	Semi-competition coils
Steering	1966 Shelby rack and pinion
Wheels	Halibrand knockoffs, 7.50x15 front, 9.50x15 rear
Tires	Dunlop racing, 4.75/11.30x15 front, 6.00/12.00x15 rear

MEASUREMENTS:

Weight	2200 pounds
Color	Pearl blue with pearl silver accent striping
Instruments	Smith
Upholstery	1966 Shelby black leather seats with black nylon carpets
Mileage	10 mpg
Turn Radius	21 feet 9 inches
Stopping	34 feet @ 30 mph
Overall Height	45 inches
Overall Length	13 feet 9 inches
Quarter-Mile Performance	10.80 @ 125 mph
Cost	$8000 (car is currently insured for $32,000)

pleted, what is left is a bunch of gaping, head-scratching onlookers watching the Cobra's "HSSSSS" rear license plate disappear into the distance.

While the ride is semi-stiff, it's not the back breaker you would assume a Ron Butler-tuned suspension would have. In fact, it's comparable to the Dodge Power Wagon we drove up to Pismo Beach. You just can't help but sit there in Cobra heaven listening to the burbling call of a well-tuned 427 as it lopes along. A pair of innocents in a Mazda pickup try to play catch-up, but after a couple of four-grand shifts, they sort of disappear from view. So does Pacific Coast Highway as we leap off her busy bustle and begin to snake up one of the semi-deserted two-lane twisters that wind their way through the Santa Monica Mountains. First we gobble up a straining VW, followed by a larger but slower garbage truck, while Jim alternately attacks and repeatedly backs off the throttle.

The power was as smooth as it was awesome, but so was the roadster's ability to negotiate 20-mph turns with the alacrity of a mountain goat. Then somehow, on a dirt-covered turnout, it was my turn to take the wheel. With a feather on my foot and discretion riding shotgun, I gingerly slipped my way onto the asphalt. My initial reaction, as I tried to coordinate the gas-to-clutch movement, was surprise at how close the clutch and brake pedals are to one another. Once you've mastered the trick of not bogging the engine or smoking the snake's skins, you can concentrate on how it stops, shifts and steers. McCashin's Cobra does all three with minimal effort. Even with a quick rack and pinion, spinning a set of wide Dunlops over which rests plenty of 427 iron, the steering effort was as if we had a power booster. Shifting was equally as easy, even though the reverse and low/second gates were extremely close. Whether you realize it or not, these Cobras featured Girling disc brakes on all four corners, with both front and rear pairs having their own master cylinder. Nonetheless, the ease of stopping approximated a power-assisted system, which is to say that even a 90-pound weakling could generate enough push to stop it on a dime.

It remains patently obvious that Shelby's Snakes were and are one of automania's last bastions of masculinity. There is nothing about them that is subtle or sedate. The wheel and tire combinations, their squatlike stance, the booming exhaust note, the lack of any creature comforts and the very fact that they have no side windows and that what passes for a top remains a joke, is strictly for the hard-core purist. And the people who own them . . . well, they are hard too. Jim carries a card that says, "Unhappiness is a Cobra being driven by a 16-year-old girl." **HR**

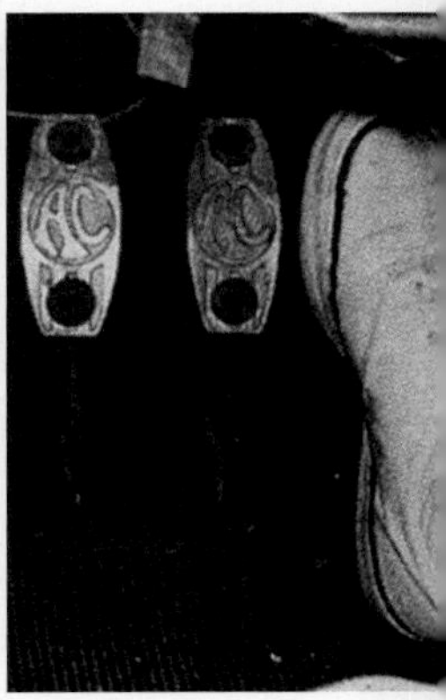

And to think you could buy one of these Filthy McNasty machines ten years ago for ten grand. Now they are selling for more than 30 Gs. Lower opening is for oil cooler. Hoop rollbar is part of the package, as are the wheels and tires. Black-chromed exhaust system is made up of two-inch pipes feeding into 36-inch-long collectors fitted with 30-inch glasspack mufflers. Dave Kent added hood scoop and more material to inside of wheel openings. While pedal detail is nice, Limeys must have tiny feet.

A scorching summer sun beat down on the black surface of the lot and reflected in shimmering waves onto the face of the gaunt Texan as he strode up to the shop. Shelby had been forced to retire from racing by a painful heart condition and was in search of help in the development of his dream.

"What's up Max?" he asked as his eyes grew accustomed to the dim light.

"Not much." Balchowski stood as he spoke and gestured toward the crew-cropped youth in the office. "Want you to meet a friend of mine. Carroll Shelby, this is Pete Brock; he just moved back to California after a stint in Detroit with GM Styling."

Balchowski was a legend in California racing and his shop was a crossroads of valuable information, but not one of the three men imagined the impact on motor racing of this coincidental meeting in the town made famous for its "discovered stars."

Brock was not Shelby's only find. He had always had good luck with people. His disarming Texas charm and droll candor made him easy to like. Only a month before the Hollywood encounter he had made another important acquaintance on the lofty, windswept flanks of Pikes Peak. While working the Ford enclave, Shelby encountered David Evans. An easy chat quickly turned to enthusiastic banter. Evans had raced against DePaolo in the 20s and 30s and had learned the art of tuning racing engines on the dusty thunder bowls and fairgrounds of the midwest. In 1958 he was selected by Donald N. Frey, Executive Engineer of Ford Division Product Planning, to head up the engine development side of Ford's new three-man team charged with the responsibility of building a racing program. Evans was one of many racing giants who would surround and populate the organization that Shelby could scarcely have imagined that brisk Fourth of July of 1961.

Only a month later Shelby challenged the new friendship. He wrote Evans in Dearborn describing his idea of building a medium-priced sports car based on the AC Ace with Ford power and requested one of Ford's new small-block engines to use in a prototype. By the time Evans called Shelby in response to the inquiry, Shelby was well along in negotiations with Charles Hurlock, owner of AC Cars at Thames Ditton on the outskirts of London. A modified Ace chassis was soon under construction.

It was November before Evans had shipped enough parts to build a complete 221 cubic inch V8 and four-speed gearbox. He had expected little of the association with Shelby so had proceeded to send parts without a word to program head Frey.

Shelby went to England and spent the winter months helping Hurlock's staff develop an integrated, reliable unit from the vintage chassis and the modern, powerful engine. After the first month the original 221 engine was replaced by the stronger 260 which Evans shipped to England as soon as it was available.

In March of 1962 CSX0001 was shipped to Santa Fe Springs, California, where Shelby had rented part of Dean Moon's shop. A second 260 had been shipped to Moon for development and when the chassis arrived sans engine it only took the talented group of hot rodders eight hours to have it running.

The following day it rumbled into life under the bright California winter sun. Shelby took his secretary/girlfriend out for a ride and pictures in the unpainted

RICK COBB

COBRA DAYTONA

By. L. C. CRANE

PHOTOGRAPHS BY DAVID FRIEDMAN

Left: Bob Holbert in CSX2287 on his way to the Cobra Daytona's first international win at Sebring in 1964. Note the spats on the rear fenders. Also the right side rocker panel looks a bit abused. Above: The four Shelby American Daytonas at LeMans 1965. A fifth car was entered by Ford of France in their colors.

aluminum roadster before it was trailered out to Riverside for some track testing.

Pete Brock was chief instructor at Shelby's racing school at Riverside by that time. He would help with the testing. Shelby had also invited Billy Krause, a young California driver, to come out for the fun. Pete and Billy took turns at the wheel. "Impressive" would understate the results. One by one the all-knowing grin of competitive men swept through the small cadre.

The second car soon arrived and was quickly completed. It went to Riverside too, but this time wearing numbers, and was driven in anger for the first time by Krause. He led running away until the rear wheel spindles broke up under the strain. Heavier frame tubes, stronger

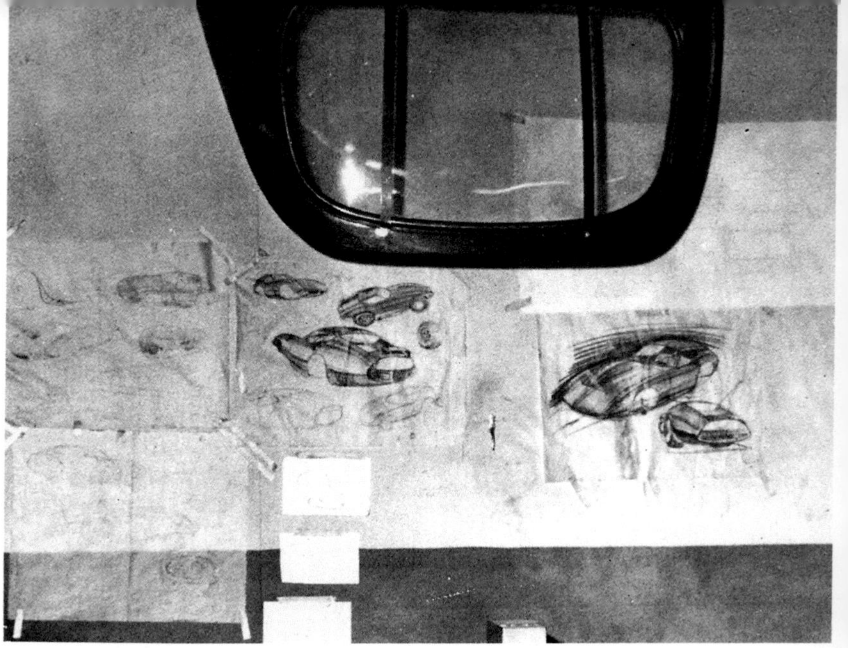

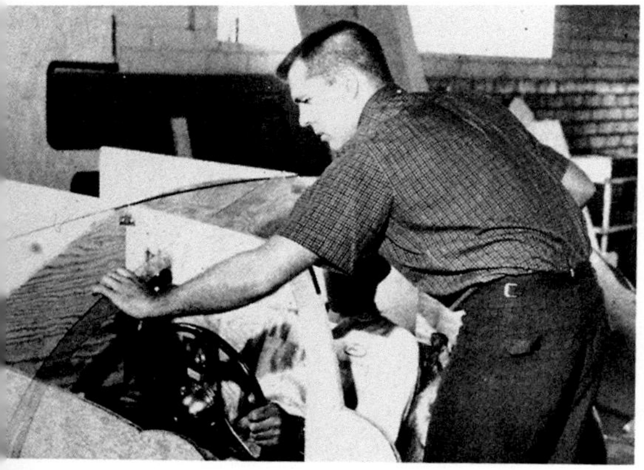

spindles, a larger radiator, stronger differential brackets, and stiffer front and rear anti-roll bars were added to the development of the first car and the Cobra came to life as the very embodiment of Shelby's dream . . . a winner.

Shelby called Evans in Dearborn and announced the success. Evans said, "Bring one back here and let's take a look." He had finally gone to Frey, who had been a life-long road racing fan and who knew of Shelby's career. He gave his enthusiastic approval. The next three Cobras were completed and shipped to Dearborn and Dan Jones' staff at Ford Engineering for *complete* evaluation and testing. Frey was satisfied and agreed to supply engines on credit until production got under way. Hurlock agreed to a similar arrangement and Shelby was in the manufacturing business.

Victories generated sales and production required larger facilities. Shelby arranged to take over Lance

Genesis. Clockwise from above: Carroll Shelby looks on as Peter Brock measures Ken Miles for a quick, new, aluminum suit. Pete finding the windshield on the buck. The concept sketches taped to the shop wall. Note the tail hoop airfoil on the center sketch and the long hardtop from a LeMans replica hanging above. Inspiration! Pete and John Olsen make a size comparison next to Pete's Falcon GT 350 sedan delivery. The buck getting skinned at California Metal Shaping.

Reventlow's large, well-equipped fabrication shop. After a brief discussion about Shelby's plans, the arrangement included the resident wizard, Phil Remington, yet another multi-talented Californian. Remington had built dry-lakes hot rods, sprint cars (with Jim Travers and Frank Coon, now Traco), sports cars for Sterling Edwards, and finally worked with Stu Hilborn to develop Hilborn's famous fuel injection system. His abilities as designer, draftsman, machinist, metalsmith, and fabricator were quite literally incredible. They were soon augmented by Ken Miles, who joined the staff as development engineer and driver. Suddenly the big shop at 1042 Princeton Drive in Venice, California was a magnet for the most gifted tuners and fabricators in the west. Ford was also present, in the person of Special Projects accountant Ray Geddes, who was sent by Don Frey to help assure a profitable product.

Dozens of cars were being produced and sold. Every weekend they would win and on Monday they would sell. They won the SCCA championship in 1962 and were well on their way in the spring of 1963 when veteran European racer and Pittsburgh, Pennsylvania, Cobra dealer Ed Hugus decided to go back to LeMans with a Cobra. Shelby agreed and had AC produce two cars. Hugus would enter CSX2142 for himself and Peter Jopp and AC would enter CS2131 for Minian Sanderson and Peter Bolton. They were fairly standard 385 hp, Weber carbureted competition cars except for their odd, elongated hardtops which were designed to reduce drag at the very high speeds required for the Sarthe circuit.

A third Ford engine ran at LeMans that year, in a diminutive aerodynamic coupe designed and built by Eric Broadley. No one recognized the fascinating little car as the inspiration that would bring millions for Ford's dollars into international motor racing, or as the harbinger of the end of the Cobras, then slightly more than a year ld.

The Lola was a bright light in Dearborn, but it did not introduce the idea of international racing to Ford. As early as 1957, Joe McKay, a junior executive selected to oversee Ford's embryo NASCAR program, sent a memo "upstairs" suggesting more international

involvement. There was no reply then, but by the summer of 1963, when Cobra power was being felt across the states and in Europe, Ford was paying attention.

Pete Brock saw an opportunity in the successful LeMans effort to fulfill one of his dreams. With Ken Miles as enthusiastic support, he approached Shelby with a proposal to build an all-new aerodynamic body to campaign in Europe in 1964. At that time Shelby appeared cool to the idea.

By September the Cobras had won another U.S. championship and Frey began listening to Shelby's interest in a European effort. In October Dearborn said "Go." Pete had recently moved back to L.A. and turned the school over to John Timanus. Shelby collared Brock and Miles and asked if they thought they could produce their car by Daytona in February of 1964. With a quick glance at each other and more than a little apprehension, they said, "Sure."

A problem appeared almost immediately. Phil Remington was an engine builder who had only seen one European race—Monaco, with the Reventlow crew in 1960. He was certain he could make his 289 cubic inch Fords beat those little 183 cubic inch Ferraris in the GT class. Shelby knew better, Brock knew better, and Miles knew better, but Remington ran the shop and his men were doing roadsters.

Fate intervened when John Olsen joined the staff. He had been working in New Zealand with Bruce McLaren. He was another jack-of-all-trades, but an outsider to the Californians, so Remington gave him to Brock. He was just the man for the job. He knew the problems of the European circuits and was excited about the project.

In late September the three protagonists rolled out the chassis of the car Skip Hudson had crashed at Daytona early that year. Ken put a piece of corrugated cardboard down for a floorboard to support a simple fiberglass seat and had a piece of steering column, with a wheel attached, welded to a small tubular structure on the chassis so he could bend it into a position he liked. Pete attached a piece of 1x4 to a wood base so it would stand vertically, to measure the minimum height of a driver. It was a tacky array of equipment that rolled out into the autumn coastal haze. Ken wriggled himself into the scruffy chassis while Pete slid his apparatus into position. With a 48-inch T-square he found the top of his coupe.

With February slightly more than 90 days away, the designer and the builders shifted into full flog. Ken drew up the plans for a triangulated superstructure to stiffen the first coupe's chassis (number CSX2287) and support the cowl and roof of the body. Pete busily reduced dozens of sketches of his aerodynamic ideas to a few finished drawings. One interesting concept was a hoop airfoil above the tail to produce down force without the drag of the ubiquitous ducktail spoiler.

When the drawings were complete, Pete proceeded to loft quarter-scale drawings of the coupe's cross sections at eleven points along its length. He then shot 35 mm slides of the drawings and projected them to full scale on half-inch plywood sheets, stood at the end of the

Clockwise from above: Ken Miles' superstructure design for the venerable Ace chassis. Note the King Cobra poster on the wall. Phil Remington welding on the cowl hoop and helping Pete with a fabrication problem. Proof that Phil did not ignore the project. Final assembly, with radiator duct and exhaust wells completed. Note steering column pyramid carrier and another coupe chassis in the background. Could it be CSX2286.

shop. One after another the projected images were traced onto the wood and cut out. Brock and Olsen attached the section and profile panels to the chassis and the coupe lived in three dimensions. As soon as it was completed the rolling buck was shipped over to California Metal Shaping to have an aluminum skin formed to its contours.

The panels were well along when Benny Howard, a famous designer of racing and corporate aircraft before the war, dropped in to see Shelby. Shel asked him to come in and take a look at Brock's drawings and give an opinion. He made some quick measurements, whipped his slide rule for a few seconds, and announced, "You might as well forget it. It'll never go. With this design, you're going to need at least 450 hp, and maybe more than that, to go 160 mph."

He and Pete discussed the roof angle for a few minutes while that worried look, which was appearing more often since Ford got into Shelby American, crept over Carroll's face. Shel took Howard to lunch. When he returned, he went directly to Brock's drawing office.

After a brief discussion of the earlier conversation Pete said, "Have I ever let you down?"

Shel allowed, "Nope," and the big gamble continued.

In less than 90 days a bulbous aluminum shape patched with plexiglass windows was pushed onto a low trailer and hauled across the coastal plain to Riverside. It had neither hoop airfoil nor duck-tail spoiler. Pete did not want the drag of the spoiler and Remington drew the line at having his men building anything as silly as an upside-down airplane wing for the back of a car. At least it was finished.

Brutal would be too strong a word to describe the shape, but sleek would also be wrong. As it idled raucously on the pit apron the unpainted skin trembled and sparkled. Miles slipped carefully into the tiny door and positioned himself in the tight seat. The rumble rose quickly to a roar as the car gathered speed and climbed out of the pits and over the bridge toward Turn Two. Ken knew Cobras as well as anyone and he knew Riverside cold. As he warmed the car the crew walked across the paddock to the back straight and brought out the watches. As the silver coupe twitched out of Turn Eight and exploded down the long straight the clocks began running: one hundred sixty-five, first time out. Suddenly dozens of teeth were exposed behind a row of grins.

Tony Hogg joined Miles for a few laps at speed in the new coupe. He described the experience in *Road & Track:* ". . . shatteringly noisy." He expanded with, "The driver works hard, due to the rather outdated suspension, the car must be driven with considerable *elan* . . ."

Pete Brock's intuition about how cars behave on the ground had been right. The coupe was still about 10 mph slower than Ferrari's GTO, but the Cobra's remarkable acceleration out of the corners would give it an advantage for lap times. It could win.

Pete took it out for a few laps too. Several tests were made on various systems. The radiator was much more efficient than the roadsters because the hot air was ducted out through the hood. The brakes ran cooler and the car not only ran faster, but was much more stable as well.

One thing was odd. The rear quarter windows were molded to form scoops to take cooling air into the rear fenders for brake cooling. The air pouring off the windshield actually passed the scoop a couple inches out from the car, forming a low-pressure area in front of the scoop. The rear tires acted as centrifugal pumps and forced the heated air *out* the quarter windows. A small plexiglass vane was devised to direct some of the air coming off the windshield into the scoop. It worked. An unexpected advantage of the coupe's shape was a 25% improvement in fuel comsumption. It was ready for its first major test in the heat of competition.

The Daytona Continental in Florida was the new car's debut. A minor problem in technical inspection was the rear tire exposure, so a triangular flare panel was fabricated and riveted onto the lower edge of the rear fender.

Dave McDonald and Bob Holbert were the drivers. After eight hours they were leading overall with a lap record of 2:08.2 at 106.989. The Series Two GTOs were far behind when Holbert pulled into the pits with an overheated differential. John Olsen, probably driven by some paternal instinct, dove to the pavement and squirmed under the tail to diagnose the problem. There were two fuel fillers, one at the top of each rear fender. The fuel poured into the left and side and air was dispelled out the right. When the tank was nearly full the fuel followed the air out the right filler and spilled over the tail and down to the red-hot differential case. The fuel exploded on contact. An alert crew member grabbed John's exposed ankle and jerked him from the conflagration as three extinguishers flooded the area with oxygen-eliminating vapors. John was seriously burned, but not critically. The car was scorched and out of the contest. The problem was the failure of a small electric pump installed to circulate the differential lube through a cooler in the tail of the body. John and the car were both functional in time for Sebring.

Early in March, before the second pilgrimage to Florida, the entire team trailered out to Riverside for pre-race testing. For further verification dozens of yarn strips were taped uniformly over the right side of the coupe's body and photographed at speed by Dave Friedman from the passenger seat of a roadster. The A-pillar vanes were again modified to catch and direct the air smoothly to the rear brake scoops rather than trip it and create a turbulent flow rearward. Ken Miles' beslotted 427 prototype roadster was also tested that day. The team returned to Venice and last-minute modifications before packing for the second cross-country trip to Florida.

On March 21, 1964, the "Daytona," as it had been christened, won its first international victory with a resounding upset of Ferrari as well as Mecom's team of GS Corvettes. Holbert and McDonald brought CSX2287 in fourth overall and won the GT category. They were followed by three roadsters. A clean sweep for Ford.

Shelby was confident enough after Daytona to bring two Ford executives to Sebring for the show. They were not disappointed and a few Ford dollars began to be diverted from Holman and Moody, by then involved with the new Ford GT program, toward Shelby and a serious season in Europe.

With a little help from his friends at Ford, a little more from Shell Oil and a lot from the Shelby American coffers the Cobras became a real threat to Ferrari's dominance of the GT championship.

The 1964 European season offered a good education to the Shelby crew. The outcome could be a sad commentary on F.I.A. politics, but the championship stands as it played, through several controversies. Ferrari won on points while the Cobra Daytona proved the Italian firm's days of dominance were numbered. Here is a brief description of Shelby American's education.

The Targa Florio on April 26 was a disaster. Three roadsters were entered and two broke. Dan Gurney and Jerry Grant finished eighth in a badly abused car, but the real European contests were still to come.

The first Daytona, CSX2287, made its European debut at the 500 Km of Spa in Belgium. Ferrari won the GT points after Phil Hill made three pit stops to remove dirt which was clogging the fuel filters. Before the finish, however, Phil had recorded the GT lap record of 4:04.5 at 129.01 mph. The coupe would not be needed at the next event in Germany and could get a thorough preparation before LeMans.

Clockwise from above: Lighting off CSX2287 for its first tests at Riverside. Ke[n] Miles rounding Turn Eight before the long straight. Note the Ford mufflers wit[h] single exhaust. Dave McDonald, Bob Holbert, and Shelby pose with CSX2287 i[n] its Daytona livery. Note the rubber bumpers for setting the nose on the groun[d]. The conflagration. You can see John Olsen on the pavement behind the fog, Mile[s] and Gurney look on from behind the hood. After Daytona came the tuft tests a[t] Riverside. Note the air still pushing out of the rear quarter window.

Three of the six roadsters entered for the Nurburgring 1000 Km race finished first, third and fourth after considerable trouble, but they won the points. The big news was the race debut of the Ford GT. It went fast, then it broke. Before LeMans the Fords would have to find reliability.

The tiny facility on Princeton Drive was working far beyond its capacity with production and SCCA racing. It would not be possible for Remington's remaining crew to produce another coupe. Even though Remington was now acutely aware of the coupe's superiority in Europe. However, they were able to modify a second chassis, number CSX2299, while Shelby contacted his long-time friend Alessandro DeTomaso regarding the construction of a body in Italy. DeTomaso recommended Carrozzeria Grand Sport in Modena.

In the pressured construction of this second chassis an odd error occurred. The cowl hoop was accidentally made two inches too tall. All the triangulation was installed and attached to it and it was shipped to Italy unchecked. The Italians, who were accustomed to inconsistent dimensions anyway, simply made the body to fit the structure—giving it two inches more head room. When the car entered competition it was the only coupe Dan Gurney could fit in, so it became his car.

Carrozzeria Grand Sport was contracted to deliver the car to Ford of France in time for final preparations for LeMans. Normal practice for Ford during the European campaigns was to use the Ford service facilities in the host country to prepare for any particular event. An attempt was also made to employ drivers of that nationality. Both coupes arrived in France and were readied for the big test.

On June 22 Bob Bondurant and Dan Gurney brought CSX2299 across the finish for the GT win and fourth overall at LeMans. It had lost its oil cooler at midnight and soldiered through with an overheating engine and eventually falling oil pressure, but it stayed ahead of the slowly gaining GTO and won. A disqualification had stopped CSX2287 when it failed to start after a refueling stop and a mechanic attached a pit battery. Both of the Ford GTs failed to finish.

After LeMans all the Fords were

trucked back to Ford of France, where they were met by Shelby engines for the Cobras and Holman & Moody engines for the Ford GTs. Only two weeks were available for reconstruction before the 12-Hour contest on the very fast Rheims circuit.

The Rheims event started at midnight. By 1 a.m. on July 4 Gurney and Bondurant were out with broken gearbox mounts in CSX2299. After six hours CSX2287 was out with the same ailment, after Ireland and Neerpasch had survived a broken exhaust pipe and its time-consuming repair. Ferrari swept the points again, but the R.A.C. Tourist Trophy on the twisty Goodwood circuit in England would give the advantage to the Cobra's torque.

Gurney won the GT points in CSX2299 at the R.A.C. Tourist Trophy on August 29 and finished third overall. Phil Hill finished well down in CSX2287 after a pit stop to repair a cracked oil line.

While the first two coupes were actively collecting points, three more cars were being prepared for the final championship round at Monza in October. One of them was to be a secret weapon. The 427 Cobra was well along in its development in the states, so Shelby decided to enter one as a prototype coupe for the high-speed Monza course. Two inches were added to the center of Daytona chassis number CSX2601 and the necessary 427 mounts welded into the structure. Upon its completion it was sent to Carrozzeria Grand Sport with instructions to lengthen the body to fit and to add a bubble on the hood to clear the higher Webers. By September the car was finished and painted and readied for shipment to Ford of Italy where the California-prepared engine and gearbox would be installed in time for Monza. It was not to be. On the way to the Tour de France the transporter carrying the third coupe, CSX2300, came to a sudden stop against a bridge abutment. The coupe left its tiedowns and mangled itself against the truck. Shelby needed at least three GT class cars to run against Ferrari on his home ground and the fourth coupe, CSX2286, was still not complete, so the word was quickly passed to Modena to cut the two inches out of CSX2601 and prepare the chassis to accept a 289 driveline. Unfortunately, after the work was done, the Monza 1000 Km was cancelled in a dispute between the organizers and the F.I.A. over the homologation of the Ferrari 250/275 LM. The Tour de France and the final championship meeting at Bridgehampton, New York, suddenly became very important to Shelby and the Cobras.

The Tour de France was for GT and Touring cars, but the Cobra troops would have been better off to take a pass. They were too inexperienced in the art of rally/racing. The two coupes that made the event were CSX2287 for Trintignant and de Saint-Auban and CSX2299 for Bondurant and Neerpasch. CSX2300, planned for Ligier and Morrough, suffered the aforementioned stoppage and was a DNS.

After a night rally the first race at Rheims on September 9 was won by CSX2299, followed by CSX2287. CSX2287 won at Rouen on September 10 and CSX2299 finished sixteenth after a throttle cable broke. A late-night rally section to LeMans finished CSX2299 since there was not time to service the cars. Trintignant brought an ailing CSX2287 in third at LeMans, but retired too before the night rally began. Ferrari's incredible reliability and experience gave them the win and the championship. Even though the roadsters swept the Bridgehampton event there were not enough points available to beat the GTOs. The season score was Ferrari 84.6 poits, Cobra 78.3.

By the end of the year five coupes were completed and all were shipped to California for a thorough preparation for the 1965 season. CSX2286 was still not complete and would not compete in the first few events. While the cars were arriving from Modena, Pete Brock was working on Shelby's ace-in-the-hole for the 1965 season.

Pete had spent several weeks of the fall completing drawings and a clay model of a 427 coupe with a potential top speed of 215 mph which would very likely bring the overall GT Prototype championship to California. Early in 1965 Peyton Cramer was sent to England on errands for Ford. On his itinerary was a stop at Radford Coachbuilders to check on the reason for the astronomical price of the GT 40 interiors they were furnishing. While there the subject of the super coupe came up. They said they would gladly produce the body for $3,000 if they could retain the very lucrative GT 40 interior contract. Cramer, always the bean-counter, jumped at the opportunity to make himself a hero with the home office, contacted Shelby with his great news. The plans and the model were shipped and Radford, with only one small coachwork project at that time, went to the street to find panel beaters.

Pete had wanted to work with the Italians because they worked as artists like he did, not from critical dimensions. Months went by with the English telling Shelby by phone that work was progressing. Finally, one of the Shelby crew, who was in England for an event, stopped by at Radford. He got on the phone to California immediately to announce they had a serious problem and that Shelby had better get Pete over there right away. When Pete arrived he was sick. The body was eight inches too wide in the middle and they had used ¼-inch steel plate and 1½-inch tubing to

For the complete story of the racing cobras, pick up a copy of THE RACING COBRAS. It was written by John Christy, who was there before the cars were built. The photographs are from the collection of Dave Friedman, who was the staff photographer and has everything the team did on film. This is the real stuff. The times, dates, places, and the names as they entered and departed the scene. A must for the racing historian.

Steven Earle

Steven Earle

Clockwise from left: The enormous 427 race engine. The first 289/427 Daytona under construction in Modena. The Type 65 coupe now finished in 1981 heading for Riverside's Turn One. The Type 65. Pete Brock and Mike Dopudja, who finished the Type 65 for Craig Sutherland. The completed 289/427 before it was cut up and made back into a 289.

support the skin. They had no idea about race car fabrication. Pete called Shelby and said they should just scrap the idea since the season was already well under way. Ford came to the support of the project because their J-Car was not working at that point. They again threatened to pull the GT 40 contract. Radford said they would hire the best race fabricators available and do the job in six weeks. The engine and chassis were already in England, so Pete planned to stay for a few weeks and watch the progress, but it couldn't happen. The English have no ready source for metal supplies and the fabricators Radford hired were the ones that couldn't get work at the real racing shops. Three months went by with little progress, so Pete called Shelby and said they should

ELLIS TAYLOR

fly everything for the super coupe to California and finish it there. Too much time had transpired by then, Ford's Mark Two was working well and the Mark Four program was beginning to show results, so Ford said "Enough." The car languished in California for a long time and was finally sold to a buyer in England where it was slowly stripped and left to deteriorate even further.

While the super coupe was being still-born, Ford was even more frustrated with Holman & Moody's inability to find reliability with their GT in spite of huge expenditures. In January of 1965 Ford pulled the GT program out of Holman & Moody and passed it to Shelby American. The announcement was made by John Wyer of Ford Advanced Vehicles in England, who was subsequently made European Operations Manager for Shelby American. Ford wanted nothing less than consistent overall wins for the Ford GT.

At Daytona on February 28, 1965, the full meaning of that edict became clear. The four Cobra coupes entered for Shelby American were to sprint with the Ferraris while the Ford circulated in reserve for the final assault. Ford got their win, but only one Daytona coupe expired. On lap 155 CSX2601 lost its engine, leaving CSX2299 in second, CSX2602 fourth behind the second Ford, and CSX2300 sixth. The legions of Ford personnel loitering in the shadows all smiled wryly and headed for the airport. There was only a month before the second test in Florida.

Steven Earle

Clockwise from the left: CSX2299 with its 1965 teammate on the European campaigns. The Shelby American race shop late in 1964 with preparations well underway for 1965. An original 1965 FIA race motor, showing the Weber 48IDA carburetors which were made for the Fords.

On March 27 the Ford steamroller hit again at Sebring. The Daytona coupes were leading at 5 p.m. when the sky blackened and opened up. The ancient airfield was awash for hours, but when the finish came at 10 p.m. it was under a clear, star-filled sky. Ford had its overall win and the Daytonas took the first four places in the GT category, with CSX2299, CSX2601, CSX2300, and CSX2602. With growing confidence the cars were loaded for the trip to Italy and the Monza 1000 Km.

Allen Mann Racing did the preparation and transportation of the two Shelby Daytonas (CSX2601 and CSX2602) and the Ford GTs for Monza on April 25. Very little GT opposition from Ferrari made a more costly team unnecessary. Ford lost to the big Ferrari prototypes, but the Daytonas finished one and two in class.

The Cobra roadsters were first, third and fourth in class at the tiny Oulton Park circuit in England on May 1. The odd Willment replica coupe appeared there for the first time.

The Targa followed Oulton Park, but Shelby opted to give it to Ferrari and save his cars. Allen Mann Racing used the time for some major service before the 500 Km race at Spa.

On May 14 Ferrari announced his team's withdrawal from competition in the GT category. His reason was valid. The C.S.I. had, correctly, refused to ratify the 250 LM for the category as only a handful had been built, but they had also withheld ratification of the 275 GTB through the first several months of the season though 116 examples were in distribution. It came as good news for the Shelby American camp, but bad news for Ford. All the SEFAC energy would be immediately poured into the prototype cars toward overall wins.

The Cobra team arrived at Spa on May 16 with perhaps just a bit too much confidence. They had underestimated the GTO privateers. Harry Sutcliff beat the Daytonas for the GT points. Bondurant finished second in CSX2601 and Whitmore withdrew CSX2602 with a driveline vibration rather than cause

JACK BRADY

any further damage to the car. Only two weeks separated the Spa event from the Nurburgring 1000 Km.

Bondurant set the tone of the May 23 Nurburgring event by setting the absolute GT record on the mountain circuit at 9:18.5 in CSX2601. A day of tire testing before the event put the Cobras on (horrors!) Firestones. Bondurant led the GT class from flag to flag with the other two Daytonas finishing second and third in class. Still ahead, however, was the biggest test of the year.

The French countryside was lush and green with American Ford dollars as the huge transporters rolled into the villages surrounding the Sarthe circuit. The two-story Ford facility was filled with crews and cars from Allen Mann Racing, Shelby American, Ford of France,

COBRA DAYTONA COUPES—THE RECORD

1964

DATE	EVENT	RACE NO.	CHASSIS NO.	FINISH POSITION	DRIVERS
February 12	Daytona Continental	14	CSX2287	DNF	Dave McDonald/Bob Holbert
March 21	Sebring 12 Hours	10	CSX2287	4th Overall/1st GT	Dave McDonald/Bob Holbert
May 17	Spa 500 Km	1	CSX2287		Phil Hill
June 20-21	LeMans 24 Hours	6	CSX2287	Disqualified	Jochen Neerpasch/Chris Amon
		5	CSX2299	4th Overall/1st GT	Dan Gurney/Bob Bondurant
July 5	Rheims 12 Hours		CSX2287	DNF (gearbox mount)	Jochen Neerpasch/Innes Ireland
			CSX2299	DNF (gearbox mount)	Dan Gurney/Bob Bondurant
August 29	R.A.C. Tourist Trophy	22	CSX2287	11th (pitted to replace oil pipe)	Phil Hill
	Goodwood	21	CSX2299	3rd Overall/1st GT	Dan Gurney
September 11-20	Tour de France		CSX2287	DNF	Maurice Trintignant/Bernard de Saint-Auban
		187	CSX2299	DNF	Bob Bondurant/Jachen Neerpasch
			CSX2300	DNS	Guy Ligier/Henri Morrough

1965

DATE	EVENT	RACE NO.	CHASSIS NO.	FINISH POSITION	DRIVERS
February 28	Dayton Continental	13	CSX2299	2nd Overall/1st GT	Jo Schlesser/Harold Keck
		14	CSX2602	4th Overall/2nd GT	Rick Muther/John Timanus
		11	CSX2300	6th Overall/3rd GT	Ed Leslie/Allen Grant
		12	CSX2601	DNF (engine)	Bob Johnson/Tom Payne
March 27	Sebring 12 Hours	15	CSX2299	4th Overall/1st GT	Jo Schlesser/Bob Bondurant
		14	CSX2601	7th Overall/2nd GT	Bob Johnson/Tom Payne
		12	CSX2300	13th Overall/3rd GT	Allen Grant/Ed Leslie
		16	CSX2602	21st Overall/4th GT	John Timanus/Rick Muther
April 25	Monza 1000 Km	48	CSX2601	8th Overall/1st GT	Bob Bondurant/Allen Grant
		49	CSX2602	9th Overall/2nd GT	Jack Sears/John Whitmore
May 16	Spa 500 Km		CSX2601	5th Overall/2nd GT	Bob Bondurant
			CSX2602	DNF (crash)	John Whitmore
May 23	Nurburgring 1000 Km	54	CSX2601	7th Overall/1st GT	Bob Bondurant/Jo Schlesser
		55	CSX2602	10th Overall/2nd GT	Jack Sears/Frank Gardner
		56	CSX	12th Overall/3rd GT	Jo Schlesser/Andre Simon (Car entered by Ford France — ran in their colors)
June 19-21	LeMans 24 Hours	10	CSX2287	DNF	Bob Johnson/Tom Payne
		9	CSX2286	DNF (clutch)	Dan Gurney/Jerry Grant
		11	CSX2299	8th Overall/2nd GT	Dick Thompson/Jack Sears
		12		DNF	Allen Grant/Jo Schlesser
		59	CSX2602	DNF	Peter Sutcliff/Peter Harper (Running in prototype class)
July 4	Rheims 12 Hours	26	CSX2601	5th Overall/1st GT	Bob Bondurant/Jo Schlesser
		27	CSX2299	9th Overall/2nd GT	Jack Sears/John Whitmore
August 15	Coppa Di Enna		CSX2601	3rd Overall/1st GT	Bob Bondurant
			CSX2299	5th Overall/2nd GT	Jack Sears

Clockwise from top left: CSX2299 with its hydraulic jacks activated during the GT record-setting performance at Daytona in 1965. The drivers were Jo Schlesser and Hal Keck. CSX2299 today, repainted but not restored. All the brake rotors are covered with radial heat cracks.

An original interior and dash layout, with the speedometer in place but the tach missing from a hole at the far left. The Sebring tech sticker from 1965.

and Dearborn's engineering staff. Veteran campaigner, John Wyer, sequestered his staff out of the frenzy at La Domaine de Beauchamp, a lovely old chateau converted into a country hotel near Chartre. Frenetic activity resulted in many changes being made to the cars, some successfully. Vast numbers of white shirts and ties followed Leo Beebe around speaking a strange midwestern tongue while Carroll Shelby wore a very long face under a terribly Fordesque crewcut. To top off the agony, cancellation of Thursday's practice precluded any rest for the crews when the practice was moved to Friday.

A brilliant sky and oppressive heat scorched the start. By mid-morning on Sunday only veteran CSX2299, driven by Dick Thompson, survived from the Ford steamroller. Everything else had blown its engine. The survivor did so without oil pressure for seven hours. The Ferraris all turned to junk as well except for N.A.R.T.'s 275 LM, which continued to run on schedule like a turbine train. A LeMans win remained in Ford's future and Shelby wondered about making chili for a living.

The official Ford dispatch after the debacle read: "The experience here demonstrates again what we knew and what we have learned from a program such as the three-year Indianapolis effort. We are prepared to go forward on this basis."

The wreckage was collected and reassembled. Shelby sent two coupes to Rheims for the July fourth contest. They motored to an uncontested GT victory and the world championship.

The Cobras ran two more small races, but they were certainly anticlimactic conclusions to a remarkable 19 months of indefatigable labor by scores of men from several countries. Ford and America had their first world manufacturer's championship. The next time you hear someone say Ford bought that championship, do suggest that person quietly step off a tall building. Ford money and expertise were pouring into their prototype GT effort, not the Cobras. Carroll Shelby, Allen Mann, Phil Remington and a handful of dedicated craftsmen took Pete Brock's creation and made that championship with their toil. Here's to American ingenuity, British persistence and buckets of B.S.&T.

Steven Earle

Shelby Cobra

HOT STUFF

No one ever makes excuses for the Cobra

STORY & PHOTOS BY JOHN LAMM

THEY WANTED $4000 for the Cobra in the showroom. It might as well have been $10,000 or $50,000 or $100,000 as far as young John Keohane was concerned. Recent medical school grads just don't have that much ready cash. On the other hand, out back behind the dealership was an AC Ace that looked a great deal like the Cobra and that cost only $1000. Keohane bought the Ace, figuring it would make do until he could afford a Cobra. He would have one eventually.

AC's Aces looked like Cobras because they were the car's immediate predecessors. At that, they were well along the line in the company's heritage, AC Cars Limited having roots running back as far as 1903. The name of the company comes from Autocarriers, which were small 3-wheel delivery trucks. These led to a passenger version called the Sociable, a sort of pre-World War I rear-engine two-person Trihawk, if you can imagine such a thing.

Control of the company was taken over by the famous S.F. Edge until the financial crash of 1929 did him in. ACs had left their mark, however, with a number of world-speed records and a victory in the 1925 Monte Carlo Rallye (trivia: becoming the only Monte Carlo-winning car to start at John O'Groats in England).

The Hurlock brothers, Charles and William, next took over AC Cars Ltd, Thames Ditton, where they built very respectable high performance cars up to World War II. Included was a handsome model called the Ace for which they claimed 90 mph—apparently that was enough to propel one of them to the winner's circle at the RAC Rally in 1933. After the war, the company found some success with a sedan, but something else was needed, a replacement model with life and vitality.

John Tojeiro, who didn't even work for AC, came up with that replacement. This clever man had built several very quick racing specials clothed with bodies that were almost exact replicas of those made for the famous Ferrari Barchettas by Carrozzeria Touring. Unable to come up with their own design, the Hurlocks offered Tojeiro a fee plus a sum-per-car if they could adapt his specials' design into the replacement model for the sedan. The designer agreed and the new AC Ace was ready, just like that.

The early Cooper race cars relied on Fiat independent suspensions at both ends, and this same principle was developed still further for the Ace. That meant a lower A-arm with a transverse leaf spring acting as the upper arm. Crude, perhaps, but very effective, especially for the cost. In the engine compartment was AC's 75-bhp 6-cylinder engine, which dated from the Depression era. However, this engine

JAN CALIFORNIA
COBRA 63

just didn't provide enough poop for the racers, so it was soon supplanted by the Bristol 2.0-liter. Another old-timer, this lovely six was designed and raced by BMW before World War II and had yet to lose its punch.

So the Ace was launched. It would go through years of continued development, the body shape evolving and the lovely Aceca coupe version being added. There were frame and suspension changes as the model moved into the Sixties. When the Bristol engine went out of production in 1959, AC chose a 2.6-liter Ford inline-6. This was, however, a baling-wire fix, and AC needed something more powerful. The company investigated the compact Buick 3.5-liter V-8—eventually the Rover engine—but there were problems with GM. Jaguar's six was too big and heavy and apparently politics ended any chance of using the lovely little V-8 sold here in the Daimler SP250. What was next?

Though the Bristol-engined models were very successful in the U.S. in Sports Car Club of America races, AC was never able to attract sufficient customers here. The Ace remained something of a mystery car for those of us lusting for automobiles for the first time. Here you are, in your late teens in the early Sixties, living in the midwest. If your hot-rodding friends knew enough to mention the name AC (fat chance), you knew about all there was to know. There were a few stories in *Road & Track* and other magazines describing the car—independent suspension, disc brakes and all that—and the ACs appeared to be very quick at the races. Truth was, we really didn't know much about AC Aces.

But we did know a great deal about Corvettes. We knew they were fast, that they had fiberglass bodies and one day soon they'd even have independent rear suspension . . . but there was something about Corvettes that didn't quite ring true. There was always the feeling we were getting a lot of overly trick exterior styling to cover too many old parts underneath. We knew the real genius of Zora Arkus-Duntov: Not only did he make these cars quick, but also he did it with parts-bin sedan pieces. We'd get our hopes up about Corvettes entered in international races, only to come away disappointed. Granted, the Corvettes often won on U.S. road-racing circuits, but that wasn't the same thing. We wanted to see them beat the world. Even the Cunningham team couldn't get a Vette to finish higher than 8th overall at Le Mans in 1960. Do you know what finished 7th the year before? An AC Ace-Bristol. Full marks to Duntov, but there

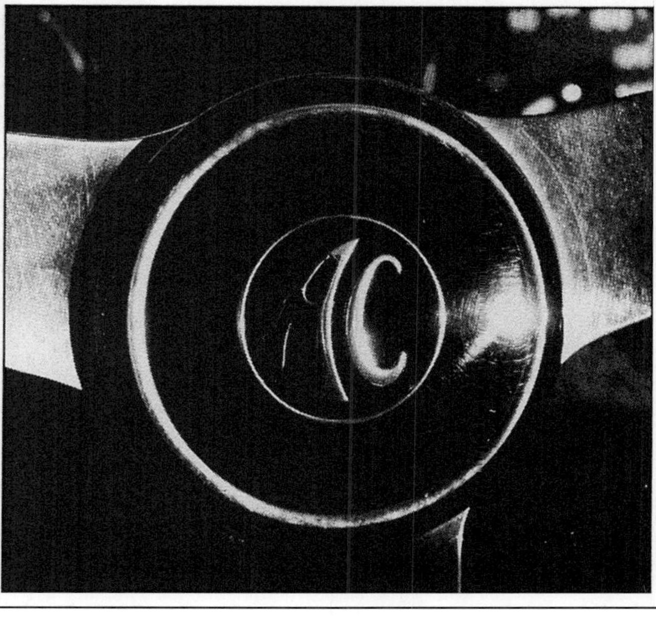

Shelby Cobra 289, provided by John Keohane, Corte Madera, California.

was something vaguely dishonest about the Corvette; sins of omission, we might say.

Then there was Carroll Shelby. Here was a man we could trust. He won Le Mans, didn't he? He wore bib overalls and a cowboy hat and talked with a drawl and stood for a lot of what we figured the U.S. could do . . . which was to win. So when we first heard about the Cobra, a sort of mystical aura began to build, and we would speculate on how wonderful it would turn out.

The Cobra. Just say it again: COBRA. So what if the chassis was English? It's not as though it was made by someone who spoke a different language than we did. And at times we figured Chevrolet did. The only thing that bothered us was that the Cobra's engine was a Ford. We hadn't really trusted Ford since it gave the T-Bird four seats. Ford owed us, and we didn't forget easily, but we could soothe the pain a bit by just saying that word again . . . Cobra.

And that's what Carroll Shelby's new car meant to many of us who were wacked on sports cars in the early Sixties. It wasn't just the sound or the color or the speed that a Cobra represented. It was the great American hope, and that was damn exciting. That was also why, a few years later, John Keohane was willing to live with a 2.6-liter Ford-powered Ace until he could buy that Cobra. And why we waited so impatiently for *that* issue of *Road & Track*. September 1962. Page 32. R&T's first Cobra test.

But it wasn't the first story about the Cobra. That had come in May 1962 in a piece in *Sports Car Graphic*, written by John Christy, with photos by Bob D'Olivo and Dean Moon. (Christy and D'Olivo would later become friends of mine when I worked at Petersen Publishing company.) What made this story doubly exciting was that it featured the very *first* Cobra, still finished in polished aluminum, with "Shelby" hand-painted on the nose and tail and the little "Powered by Ford" badge on each side. Even the engine was exciting; it was called the XHP-260, the X standing for experimental. Great stuff. Ford claimed 260 bhp, one horsepower per cubic inch, and we were nearly ready to forgive it for the T-Bird.

Then came the R&T story, confirming that the Cobra was everything we hoped for. Acceleration to 60 mph came up in a blindingly quick 4.2 seconds. That number would climb past 6 seconds in later tests of full production cars, and the top speed would drop from 153 mph to 139. But, what the hell, the car was so fast compared with anything we were driving, we didn't care.

That Ford V-8 and its Borg-Warner T-10 4-speed gearbox were the two major alterations that turned the Ace into the Cobra, though there were hundreds of detail changes. Though the tubing used was larger than that in the AC, the frame remained a ladder type, surrounded by an aluminum body. The body didn't appear much changed from that of the 2.6-liter Aces, though the fenders were flared just a bit for the larger 7.35-15 tires of the Cobra. Seen overall, however, the

ROAD TEST

AC FORD COBRA

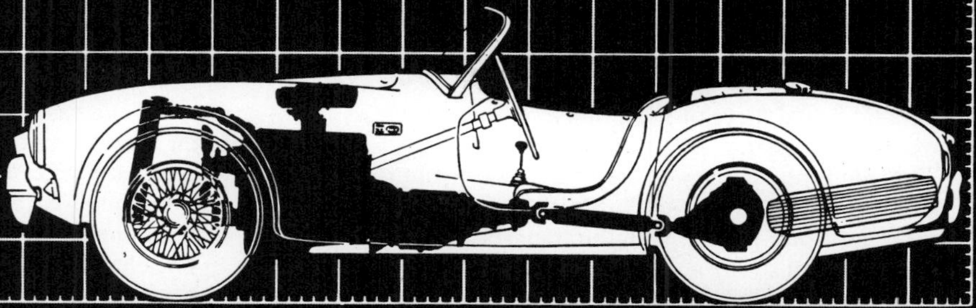

DIMENSIONS

Wheelbase, in	90.0
Tread, f and r	51.5/52.5
Over-all length, in	151.5
width	61.0
height	49.0
equivalent vol, cu ft	262
Frontal area, sq ft	16.6
Ground clearance, in	7.0
Steering ratio, o/a	n.a.
turns, lock to lock	2.0
turning circle, ft	34
Hip room, front	2x16.5
Hip room, rear	n.a.
Pedal to seat back, max	40.0
Floor to ground	10.5

CALCULATED DATA

Lb/hp (test wt)	9.6
Cu ft/ ton mile	175.2
Mph/1000 rpm (4th)	21.8
Engine revs/mile	2745
Piston travel, ft/mile	1315
Rpm @ 2500 ft/min	5230
equivalent mph	114.3
R&T wear index	36.1

SPECIFICATIONS

List price	$5995
Curb weight, lb	2020
Test weight	2355
distribution, %	48/52
Tire size	6.50/6.70-15
Brake swept area (est)	580
Engine type	V-8, ohv
Bore & stroke	3.80 x 2.87
Displacement, cc	4261
cu in	260
Compression ratio	9.2
Bhp @ rpm	260 @ 5800
equivalent mph	127
Torque, lb-ft	269 @ 4500
equivalent mph	98

GEAR RATIOS

4th (1.00)	3.54
3rd (1.41)	4.99
2nd (1.78)	6.30
1st (2.36)	8.36

SPEEDOMETER ERROR

30 mph	actual, n.a.
60 mph	n.a.

PERFORMANCE

Top speed (7000), mph	153
best timed run	n.a.
3rd (7200)	112
2nd (7200)	89
1st (7200)	67

FUEL CONSUMPTION

Normal range, mpg	n.a.

ACCELERATION

0-30 mph, sec	1.8
0-40	2.5
0-50	3.3
0-60	4.2
0-70	5.4
0-80	6.8
0-100	10.8
Standing ¼ mile	13.8
speed at end	112

TAPLEY DATA

4th, lb/ton @ mph	off scale
3rd	off scale
2nd	off scale
Total drag at 60 mph, lb	115

ENGINE SPEED IN GEARS

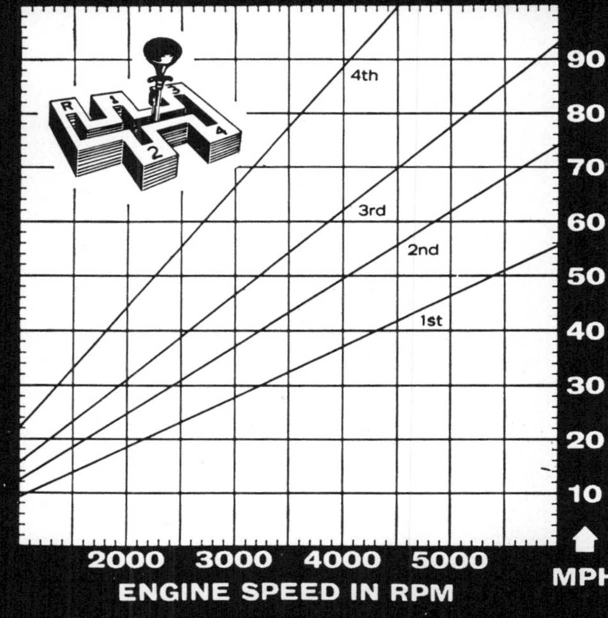

ACCELERATION & COASTING

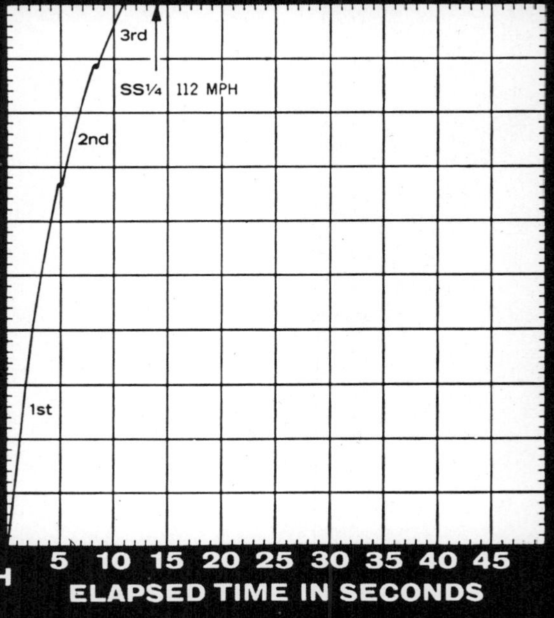

Reprinted from Road & Track, September 1962.

Cobra's look *was* different. R&T's road test report of June 1964 written, I believe, by our late friend, Tony Hogg, who later became a Cobra owner, sums it up best: "One does not really appreciate what Shelby has done to the AC design until one sees an AC Bristol and compares it to a Cobra. The AC Bristol is a nice, lean, graceful car, and the Cobra is a squat, mean and brutal piece of machinery, but nevertheless extremely handsome."

Though in practice it was lightyears away from that original Fiat-inspired suspension, the same basics still applied. In front was a lower A-arm, with the transverse leaf spring acting as both a spring and an upper arm. The same basic design continued in back, with tube shocks doing the damping at both ends. Naturally, any beefing-up of suspension or drivetrain pieces to handle the increased horsepower and torque of the Ford V-8 was done. Disc brakes were used at each wheel.

Many of those changes came about quickly because one of the first things Shelby did was to race the Cobras. Living in the midwest and waiting for the reports of those first Cobra/Corvette duels was like waiting for news from the front. We weren't disappointed. It got even better as the 289 Cobra evolved into the FIA car, then the 427s and the Daytona coupes . . . but those are three other stories.

The interior of the car came in for minimal changes from Ace to Cobra. The dashboard remained a flat panel with a glovebox on the right and neat white-on-black instruments to the left. Speedometer (ready to register suitably higher speeds) and tach were directly in front of the wood-rim steering wheel. The remaining instruments, including oil pressure and temperature, coolant temperature, ammeter, fuel level and a clock, were grouped just to the right. The leather-covered seats had enough adjustment to allow for a six-footer, but not much more. There was a top, though like all the British tops of the era, it was just to tide you over to better weather. Ditto with the side curtains.

Dr Keohane doesn't even own side curtains for his Cobra. Then again, he puts the top up only when he has to, mostly just for the winter. Such is the advantage of living in the town of Corte Madera, just north of the Golden Gate Bridge. Granted, he has to drive only a few miles to work each day, but if there's any chance to do it in the Cobra, he does. The good doctor, who serves as an orthopedic surgeon for the U.S. women's ski team, has put some 10,000 miles on the car. And he has every intention of adding many more.

The chassis number of Keohane's Cobra is 2063, and it was originally built with a 260 engine. The car was damaged in a fire, and when it was rebuilt, it was updated, making the V-8 a 289 and adding the front fenders with the vents aft of the wheels. This is not a show car, but a Cobra used and appreciated daily. Remember how long Keohane had waited for this car. He hasn't been disappointed.

Then again, I can't think of anyone who's been disappointed when he finally got to drive a Cobra. Yes, the ride is a bit rough and there's nothing supple about the way it handles bumps, dips or oversteer, but then no one ever promised otherwise. It has its rough edges, but they aren't important because you're going so quickly and having such a grand time. Perhaps it's the acceleration or the top speed or just the way the entire program came about, but no one ever makes excuses for the Cobra. What you see is what you get . . . no pretenses. That was the difference between the Cobra and the Corvettes of the time.

It's possible that no large carmaker, GM or Ford, could ever build a car like the Cobra. Committees can't create such wonderful machines. Besides, there's no money in it . . . at least not for large corporations.

One man can make a difference. Thank you, Shel.

COBRA PRODUCTION TOTALS*

1962-1965 Cobra

260 cid cars	75
289 cid cars	579
TOTAL	654

1965-1967 Cobra

427, 428 cid cars	348
TOTAL	348

GRAND TOTAL 1,002

**Breakdown of street vs. competition cars detailed in respective chapters.*

260-289 COBRA
a star is born

The Cobra was Carroll Shelby's dream car. And Carroll Shelby — and maybe *only* Carroll Shelby — could have made it happen. Granted, he came along at precisely the right time, but he followed through. When opportunity knocks, you don't want to be out back in the outhouse.

Shelby always had a certain 'image' of his dream car. It was probably more an idea of what it would do, rather than how it would look or how it would be produced. Those things could always be worked out later. To his everlasting credit, and although in the beginning he was not dealing from a position of strength, Shelby steadfastly refused to compromise. As he said in 1964, "I built the Cobra with my heart, rather than. . . for a profit."

Carroll Shelby always had the desire to build a sports car of his own. Like a thread, that goal winds through his life, often rising to the surface briefly. It was the major reason why he decided to race in Europe — to see how limited production manufacturers operated. He almost started building a prototype two or three times during his racing career and when health problems forced his retirement in 1960, he made up his mind.

Before his retirement — in 1957 to be exact — Shelby tried to interest General Motors in his idea of building a small, lightweight sports car based on a European or Japanese-built chassis mated to an American small-block V-8 engine. In 1957, all GM could see were Corvettes. Shelby had driven almost every sports car around by the time he retired and when he learned that the Bristol Aeroplane Company had ceased production of Bristol engines for AC Cars, Ltd.'s sports cars, he recalled his experiences behind the wheel of AC Bristols. He knew they had potential. About that same time, he learned of Ford's new 221 cubic-inch small block V-8. In his best wheeler-dealer style, he succeeded in interesting both parties. A short time later, the very first AC 'Cobra' arrived, engineless, at the Los Angeles airport. And just eight hours after it was unloaded Shelby, Dean Moon and Moon's staff had it running! Everyone realized that the car had a long way to go, but Shelby knew that this was the car that he had been dreaming about.

Cobra chassis and frames were built at AC Cars, Ltd., outside London, England. The chassis and frame parts were welded together, suspension components were attached and the body panels — built for AC by local subcontractors — were assembled and attached to the frame. Engines and transmissions were shipped by Ford to Shelby American

Pete Brock road tests an early 260 Cobra. Brock initially came to work for Shelby as an instructor at the driving school.

AC Cars, Ltd.'s assembly line. After chassis and subframes were welded together, suspension, steering and brake components were installed.

Firewall and footboxes installed, the cars await body skin, inner panels and electrics before receiving paint and hardware.

Inner panels were pop-riveted to subframe. Body was attached in the same manner. Chassis and subframe assemblies were painted black.

After paint and interiors, cars were trucked to the London docks for shipment by freighter to Los Angeles.

Once in Shelby American's Venice plant, Cobras received engines, transmissions and driveshafts as well as final detailing. Each car was road-tested before shipping to the dealer.

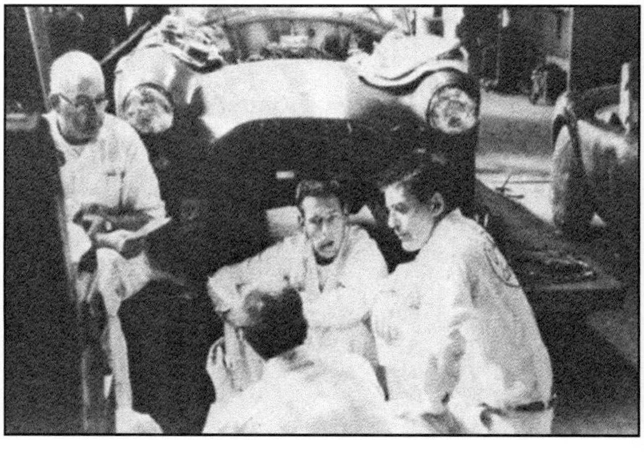

Some touch-up painting was required because the crews of the freighters bringing the cars from England had the habit of walking over the cars once they were in the ships' holds.

and installed at Shelby's plant, initially in Santa Fe Springs, California. (In order to get some cars to dealers on the east coast, a few of the early cars were completed at soon to be Shelby dealers; Ed Hugus' Ford dealership in Pittsburgh, Pennsylvania and Tasca Ford in East Providence, Rhode Island both received a few of the earliest cars to complete.)

Each Cobra received a Shelby American serial number as soon as its construction was begun at AC. Each number began with a prefix; if the car was intended for export (to the United States), the prefix was 'CSX' — which stood for 'Carroll Shelby Export.' Cobras that remained in the United Kingdom or Europe had 'COX' (Cobras exported to Europe) or 'COB' (Cobras sold in Britain). Each Cobra serial number had a four-digit number after the prefix. The first number was either a 2 or a 3, depending upon whether the car was a leaf spring car ('2') or the later coil spring model ('3'). The last three numbers represented the actual consecutive production number. The only deviation from this was the very first car. It was given the serial number, 'CSX2000.' Shelby told everyone that the 'CSX' on that car stood for 'Carroll Shelby Experimental.' Subsequent cars began with 2001, 2002, 2003, etc. The majority of Cobras exported to the United States were shipped by boat. The exceptions were certain race cars that were needed quickly; these arrived by air.

The leaf spring Cobra chassis was actually little more than two 3-inch diameter parallel tubes that were connected, ladder-style, by crossmembers and had spring towers constructed on each end. The main chassis tubes were placed in frame jigs and the crossmembers and spring towers were arc welded to them, forming a solid assembly. Next, body framing and support brackets were added to the main tubes by gas welding. The chassis and framing were built from mild steel, in a mixed framework of round and square tubing, which contributed to the car's structural stiffness. Suspension parts were assembled and attached to the chassis, followed by the steering mechanism, foot boxes, gas tank, fuel and brake lines and the wiring harness. The body panels were assembled by welding the pieces together and the completed body was installed over the frame. Lastly, the cars were painted, interiors were installed and final detailing was carried out.

At the beginning of production, AC Cars could turn out about fifteen Cobras a week but being a small quantity manufacturer, no quota was imposed; the cars were finished as they were finished. Period. Cobras were limited production cars, and as such, changes were made constantly. As a result, very few of the earlier cars were exactly alike. When a particular problem arose, it was immediately corrected and the change was carried forward on all future cars. In this manner, various 'series' of Cobras were built unintentionally. References were made to 'rack and pinion' cars, 'worm and sector' cars, 'Lucas electrical system' cars, and the like. These descriptions refer to the specific type of parts that were used in the production of that car, and others with similar parts — all of which had consecutive serial numbers. These series all overlap, initially, but as the cars got sorted out, the running changes came less frequently. At the right is a brief table of series.

The body of the Cobra was handformed from sheets of aluminum .045 inch thick. The entire body skin weighed

CSX2000-CSX2200	— 'first contract' for 100 cars; Dec. '62-April '63
CSX2201-CSX2589	— 'second contract' remaining leaf spring cars built; Nov. '63-Nov. '64
CSX3001-CSX3053	— coil spring race car production; begun Feb. '65
CSX3101-CSX3201	— 'first coil spring contract' for production cars; 1966
CSX3202-CSX3358	— 'second coil spring contract' for production cars; 1967
CSX2000-CSX2030	— completed at Shelby American plant in Santa Fe Springs (a few cars completed at Ed Hugus & Tasca Ford)
CSX2031-CSX2589	— completed at Shelby American plant in Venice
CSX3001-CSX3358	— completed at Shelby American plant at Los Angeles airport
CSX2000-CSX2074	— built with 260 cubic-inch engine
CSX2075-CSX2589	— built with 289 cubic-inch engine
CSX2000-CSX2069	— built with 3.54 rear end ratio
CSX2070-CSX2589	— built with 3.77 rear end ratio
All European Cobras built with 3.54 rear end ratio	
CSX2000-CSX2200**	— built with Lucas electrical system
CSX2201-CSX2589**	— built with Ford electrical system
CSX2000-CSX2070	— built with Lucas generators
CSX2071-CSX2200	— built with Ford generators
CSX2201-CSX2589	— built with Ford alternators
CSX2000-CSX2125	— built with worm and sector steering, flat steering wheel
CSX2126-CSX2589	— built with rack and pinion steering, dished steering wheel
CSX2000-CSX2200**	— built with Smiths gauges
CSX2201-CSX2589**	— built with Stewart Warner gauges
CSX2000-CSX2187	— built with manually adjustable handbrake
CSX2188-CSX2589	— built with self-adjusting handbrake
CSX2000-CSX2164	— built with single brake master cylinder
CSX2165-CSX2589	— built with twin master cylinder and dual reservoirs
CSX2000-CSX2149**	— built without side vents
CSX2150-CSX2589**	— built with side vents (serial number approximate; actual transition car unknown)

***number approximate; actual numbers would require physical inspection of each car.*

Cobra interior was very straight-forward and very British. Seats and dash were covered in real leather.

only fifty pounds. The car's lines were classical English roadster, based on AC's Ace. Shelby instructed AC to make a few refinements that had the effect of smoothing out the overall shape — notably around the grille and lower front section and the fender openings. The body formed a cowl that became the leading edge of the dashboard, its lines uninterrupted by the placement of the windshield.

The Cobra's instrument panel featured full instrumentation, which included a 0-8000 rpm tachometer (early Cobras had Rotuna tachs, manufactured by Faria or Smiths 10,000 rpm generator-driven mechanical tachs while later cars used Stewart Warner units) and a 0-160 mph speedometer, left and right, respectively, of the steering column directly in front of the driver. In the center of the dashboard were smaller gauges, set in a three-two-one pattern, top to bottom. They included water temperature, oil temperature, ammeter, oil pressure, fuel and an electric clock (early cars used Smiths clocks while later cars used the clock out of a 1962 Ford Galaxie). Light and wiper switches were from an early Falcon (early cars had English-made Lucas knobs). The bucket seats were finished in real leather and the steering wheel was aluminum with a wood trim. Interior panels that formed the transmission hump and the panel behind the seats were aluminum and the cars all received 'wall to wall' carpeting.

The heart of the Cobra was what throbbed under its hood, or bonnet, as the English said. Ford originally provided Shelby with 221 cubic-inch engines and on the basis of early dyno-testing, these engines — in various configurations — clearly established their high potentials. By the time that CSX2000 was completed and shipped to the States (minus engine and transmission) the 221 had been updated to 260: the bore was simply increased from 3.500″ to 3.800″. Shelby added solid lifters, a heartier-than-normal camshaft and a four-barrel carburetor to bring the output up to a respectable 260 horsepower.

The small block Ford engine was originally designed for use in the medium-sized Fairlane and as an option for the Falcon and Mercury compacts. The major elements that were considered in its design were compactness and lightness. The complete engine, with accessories, weighed 490 lbs. — about 160 pounds lighter than the big block 292-332-352 series engine — while putting out the same horsepower. Thin-wall casting techniques were employed to maintain low weight. The small block Ford had the shortest stroke of any American V-8 on the market at that time. This short stroke contributed to the reduction of weight, because shorter connecting rods could be used and a reduction in the overall size of the cylinder block casting resulted. The high bore-to-stroke ratio offered other advantages which included low piston speed, reduced frictional losses and the opportunity to use large valves which promoted good breathing characteristics.

After the first 75 Cobras were produced, Ford began offering a 289 cubic-inch version (again the result of a larger bore: 4.000″ vs. the 260's 3.800″). This engine was immediately pressed into service for the Cobras. In fact, many 260 Cobra owners, upon hearing about this 'new' engine, were so entranced by the prospect of those extra 29 cubic inches that they brought their snakes back to Shelby American for a venom-transplant. The 289 Cobra engine was the 271 horsepower high performance version, with solid lifters, higher-lift camshaft and a 460 cfm Autolite carburetor (although some cars received Holley carburetors). Cooling on the Cobras was aided by an electric fan — mounted behind the radiator in early cars and in front of the radiator in the later ones. It was thermostatically-controlled and activated when the engine's water temperature reached 90-92° Centigrade. Initially, Corvette aluminum radiators, header tanks and brackets from '61-'62 Corvettes were used, but eventually (after the first 135 or so cars), Ford supplied brass radiators and header tanks to Shelby built by McCord. Early cars' exhausts ended in front of the rear wheels. Later cars' exhausts were routed over the rear axles and ended under the rear bumper.

Final detailing included a general clean-up, last-minute inspection and the addition of anything special that was ordered by the dealer or the customer; this 260 Cobra received a custom center stripe and a racing mirror.

All-out competition engines were eventually groomed to produce over 380 honest horsepower without sacrificing reliability. While extensively modified, these engines still utilized stock Ford high performance connecting rods and distributors. The pistons and oil pumps were reworked by Shelby's competition department and special racing clutch and flywheel assemblies were used. Horsepower figures were obtained on Shelby American's dyno, but strength and longevity were proven on the toughest testing grounds that there were — the racing circuits of the world. The 289 small block Ford engine is thought by many to be the best engine ever built. It was certainly instrumental in Shelby American's World Manufacturers Championship victory.

Carroll Shelby's true goal in building the Cobra was to produce a winning race car — one that would outrun the Corvette and Ferrari production cars. He never wanted to build a lot of cars and make a lot of money. He just wanted to go racing. At the outset, the goal was to produce only 100 cars a year, because that was the minimum number that was required to get the cars certified as production cars. (This was the reason why the first contract was for 100 cars.) Shelby also knew that there was no way that he was going to sell 100 race cars a year, so the Cobras were sold as street cars. They were, in effect, detuned race cars. All options that were offered were taken directly from the competition cars. Cobras were over-engineered to accept more drive line and handling stresses than 90% of their owners would ever give them.

After the first few cars were completed, Shelby decided to see what one would actually do on a race course. In October, 1962, at Riverside Raceway in southern California, the Cobra began what was to become a racing legend. It didn't happen all at once, but those who were there didn't need to be hit over the head with a brick to realize what was about to take place. Shelby sweet-talked CSX2002 into the 'Experimental Production' class and driver Billy Krause immediately treated the Corvette drivers to a 'taillight display.' The red #98 Cobra opened up a mile-and-a-half lead before a stub axle broke, forcing the car out. But the Cobra's potential had been amply demonstrated; when the race results were reported, the big news was not the Corvette's victory but the Cobra's red-hot performance.

CSX2002, in racing trim, differed very little from the actual production cars. In place of the windshield was a small racing wind screen. The soon-to-be familiar single-loop roll bar with forward brace was installed, along with a larger gas tank. The egg crate grille was removed, front brake cooling ducts were installed under the front gravel pan and the hood sported a small air scoop and five rows of louvers across the back to aid in expelling engine compartment heat. A single Raydyot aluminum bullet-shaped racing mirror was installed on the center of the cowl, a quick-fill fuel cap was added. Koni shock absorbers were installed. The engine received little more than a complete balancing and blueprinting and Spaulding ignition, hi-rise intake with a Holley carburetor and an extra capacity oil pan. Special exhaust headers straight exhausts ending just forward of the rear wheels completed the picture.

As the Cobra began to make a name for itself in local production racing, suspension and chassis inadequacies began to show up. Each failure was corrected and all subsequent cars were updated to reflect these corrections. All of the problems that were associated with the early cars — such as stub shaft and stub axle failure, insufficient radiators, poor cooling and the imprecise worm and sector steering — were overcome by the time that the first 150 cars had been built. Strengthened parts, larger radiators, side vents, cooling ducts and rack and pinion steering all helped to solve these unanticipated problems.

After the switch to the 289 engine and rack and pinion steering, the major differences in street Cobras were in the form of which specific options each individual buyer ordered on his car. A lift-off fiberglass hardtop was available, as were different styles of wheels. All small block Cobras had wire wheels as original equipment. The early cars had 5-inch wide, 15″ diameter 72-spoke rims. They came painted in silver enamel, but chrome was an option. These cars carried 6.00/6.40 Goodyear (Shelby was a Goodyear distributor, remember?) "Blue Streak" Sports Car Special tires. As tire technology improved, a larger rim width was incorporated. Again, silver was standard and chrome optional ($150); they were 6″ wide and mounted Goodyear G-87.35 x 15 low profile tires. These 6″ rims are

CSX2002, the first competition Cobra, was raced at Riverside for the first time, driven by Billy Krause. The date was October, 1962.CSX2002 was raced at Daytona and Sebring in 1963 and then sold.

"XP" was the special Experimental Production *class; the required 100 cars had yet to be built. Krause proved that the Cobra could be a winner, as soon as the wrinkles were ironed out. It didn't take long.*

easily identifiable because of a more pronounced space around the outer end of the spokes. In a few instances, optional mag-type street wheels were available. Shelby offered American Racing 5-spoke aluminum alloy wheels, 15″ in diameter and available in different rim widths. They utilized a steel spine adaptor which bolted to the wheel using a Chevrolet bolt pattern. These wheels are fairly rare by virtue of the fact that most buyers felt the wire wheels were both sufficient and attractive.

As previously stated, virtually all of the competition parts were available as either options or over-the-counter parts. Thus, the appearance of hood scoops, fender flares, roll bars and the like — all legitimate optional equipment. Late in the model run a C-4 automatic transmission was made available; approximately 30 Cobras were built with this transmission.

Never one to shirk at gilding the lily, in 1965, Shelby offered a special model Cobra equipped for autocrosses, slaloms and gymkhanas. Called the 'Slalom Snake' and costing a hefty $1,000 more, it included a handling package of sorts: front and rear sway bars, Koni shocks, magnesium wheels and Goodyear Bluestreak tires. A single-loop, forward-braced roll bar, front brake scoops and hood scoop also were added. To decrease weight, the windwings, sun visors, front and rear bumpers and luggage rack were deleted. Slalom Snakes were offered with a matching center stripe that was color-coded to the interior color. The actual number that was sold is not known, but like most other special model Cobras, you can bet it was small.

In the fall of 1963, another special model of the Cobra appeared. It was called the 'Dragonsnake' and as its name implied, competed at the drags. Its debut was the 1963 Indy Nationals where, with dual quads and narrow wire wheels, it was driven by *Motor Trend* editor Jim Wright. It was tuned by Louis Unser of Shelby's Race Dept. and fell victim to a broken axle. The Dragonsnake came about when a few young Shelby American employees, your basic southern California hot-rodders, convinced their boss that a properly set-up Cobra would be a sure winner at the strip. He agreed to supply a car and they supplied the time and effort. The result was the 'Dragonsnake' model — like all other Shelby options and trick parts — available as a complete car ($8,250, FOB Venice, California), or as individual parts (available from Shelby American or your local Cobra dealer). The first Dragonsnake, originally red and later painted a heavily metalflaked light blue, was used to

At the end of 289 Cobra production, between serial numbers CSX2536 and CSX2589, approximately 30 cars were built with automatic transmissions. A note on the back of this photo made by Pete Brock identifies this car as the first one that was completed. Dealer cost of the automatic transmission was $185.00.

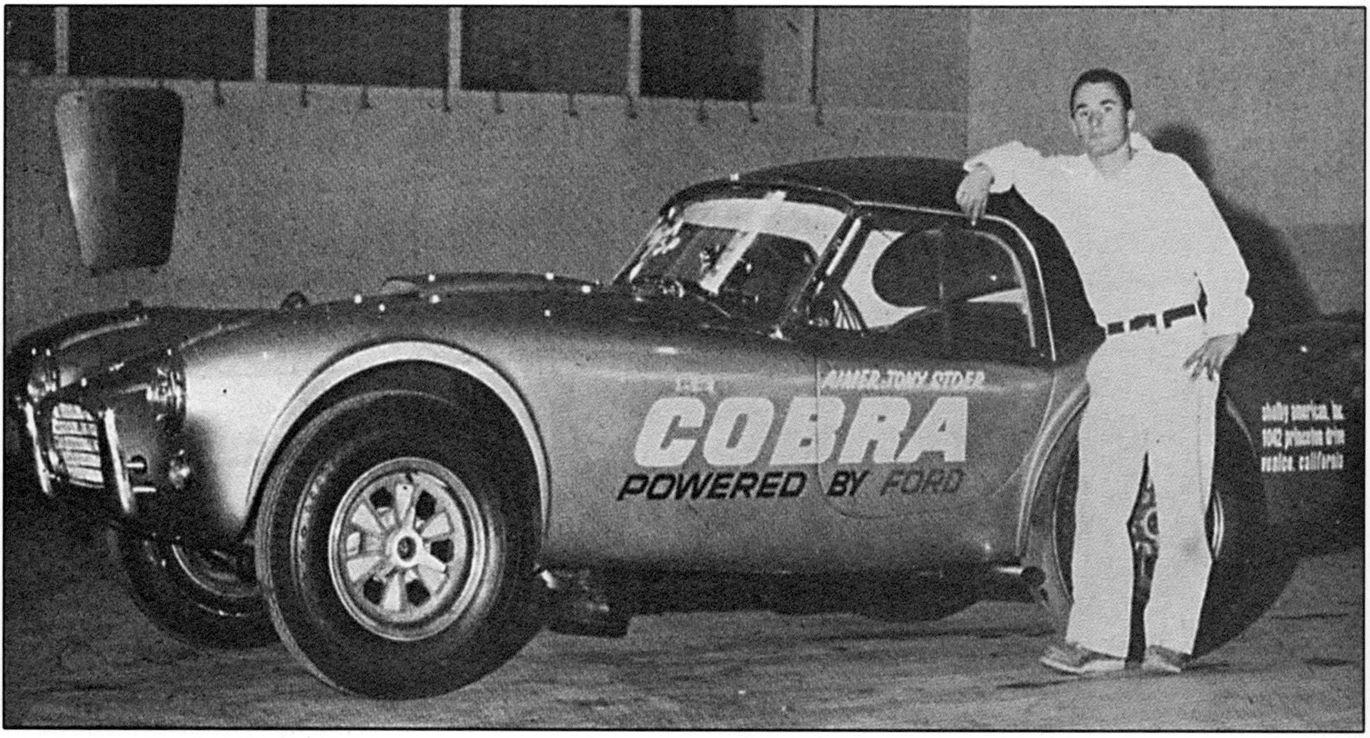

The first Dragonsnake was CSX2036. Shelby American employee Tony Stoer was the driver. This car used the road race "Nassau" headers without side pipes. As per NHRA regulations, standard exhaust had to be hung underneath the car, although not hooked up. Stoer's height — and attendant lack of weight — was a definite asset.

Stoer could barely see over the steering wheel, but that did not hinder the car's performance. Only a few other Dragonsnakes were actually built but because all parts were available from Shelby American, individual owners could — and did — build their own drag race Cobras.

develop drag race and acceleration packages that were immediately made available to Cobra owners who were quarter-mile oriented.

Four stages of Dragonsnake engines were developed. Stage I-D was the standard 271 horsepower 289 hi-po. When fitted with two 4-bbl. carburetors and rated at 300 horsepower, it became the Stage II-D. Stage III-D utilized four Weber 48mm IDA carburetors and yielded 325 horsepower. The full-tilt Stage IV-D Dragonsnake engine, rated at a whopping 380 horsepower, also ran the Webers but they were mated to fully reworked heads and a wild 'acceleration' camshaft. All engines were, of course, fully balanced and blueprinted and sported 'Cobra' aluminum valve covers and oil pan. Special drag race Belanger headers, terminating under the doors, were used, as was a Cobra scattershield and Ford Qk-1 aluminum transmission. The standard rear end ratio was 4.89:1, but others were available.

All manner of wheels worked, but the original Dragonsnake was shod with 6-spoke Halibrand FIA mags. Cure-Ride drag racing 90/10 uplock and 50/50 downlock shock absorbers were used, front and rear, respectively. The front end geometry was modified to produce more favorable results by de-arching and lengthening the front spring and lengthening the rear spring. A fiberglass hardtop, hood scoop, reradiused and flared rear wheel openings and a dash-mounted Sun tachometer were installed, as were competition seat belts and a few forward-braced chrome roll bars. Dragonsnakes (only a few were ever sold as complete cars) and privately owned and prepared drag Cobras had a monopoly on the production sports car classes until 1967, when—in 1964 form—they were no longer competitive.

Cobras competed in every conceivable type of competition, from Bonneville to Pike's Peak; from hillclimbs, international rallies and autocrosses to full-bore production road racing, and it was here that their name was permanently etched into automotive history. Although their victories have been worldwide, they started locally, in southern California regional SCCA events and as their victories increased, so did the expansion of Cobra racing activities: SCCA Divisional and National races, the United States Road Racing Championship series and fully-sanctioned FIA events. The SCCA victories came fairly easily as competition came in the form of independents and sponsored 'amateurs.' 1963 was the first year of the USRRC series, which had divisions for both manufacturers and drivers. The USRRC was the forerunner of the Can-Am series and was, essentially, open to all closed-wheel sports and sports-racing cars. After some early mechanical problems, the Cobras began their franchise on victory lane. They won the championship in 1963 and 1964, often beating faster and more sophisticated pure-racing machinery.

When the second Shelby American factory Dragonsnake was built from a rack-and-pinion car (serial number unknown), the first one was sold. The second generation Dragonsnake was painted a heavily metalflaked blue. Runs of 11.80 sec. @ 116 MPH were not uncommon.

CSX2128 was one of the first rack-and-pinion race cars and it made its debut at Sebring. Following that it was raced in four USRRC events as a team car and then it was sold to Coventry Motors, a Cobra dealer in Northern California. Allen Grant, who had started working at Shelby American as a welder when Shelby turned him down as a driver, became Production Control manager and interested Coventry Motors in the race car — provided he was the driver. To sweeten the deal, he happened to mention his friendship with Shelby engine builder Ole Olsen and the probability that they could field a car that was faster than the factory cars. Coventry bit. Grant and Olsen campaigned the car for six months and Shelby offered Grant a seat in a factory Cobra the following season. That was the good news. The bad news was that Grant received his draft notice. He joined the reserves and six months later was knocking on Shelby's door again, looking for that Cobra ride. CSX2128 was returned to Shelby American, updated and raced in the 1964 USRRC series before it was sold again. While Coventry had the car, it was painted bright yellow with black trim. The entire rig was a sight to behold. Pictured here are Olsen (left) and Grant.

The FIA World Manufacturers Championship was really the apple of Carroll Shelby's eye. Ferrari had held a long-term lease on the GT class, winning it every year for almost as far back as anyone could remember. Shelby's competition Cobras entered FIA races in 1963 — Daytona, Sebring and LeMans. They immediately distinguished themselves as viciously fast, but they failed to go the distance. Mechanical ills plagued the cars and Shelby found out, very quickly, that FIA GT racing was hardball. Finally, in September of 1963, Dan Gurney, driving a Cobra competition roadster, CSX2137, became the very first driver to win an international race in an American car. The place was Bridgehampton, New York.

The Cobra was an obvious winner: the chemistry was right. The basic engine/chassis combination, refined over the past year of competition, proved that the championship was within reach. The determining factor in Europe would be sustained, high speed running on the longer tracks where top speed was a deciding factor. The Cobra roadsters had the aerodynamic characteristics of a shoe box, so it became apparent that a more aerodynamically-clean body would be the key. It would yield the twin benefits of higher top speed and increased fuel economy. FIA Grand Touring rules specified that a minimum of 100 cars had to be built in a twelve month period, but the exact body style was open as long as the chassis and engine remained the same. In the fall of 1963, a coupe body was being planned. It premiered at Daytona in February of 1964 and was immediately called the 'Daytona Coupe.' It was to become Shelby's trump card. Along with the roadsters, it would bring the elusive FIA World Manufacturers Championship to the United States.

There were several distinctly different types of Cobra competition cars produced. Each type was built within a particular time frame and constructed to fit within certain rules. Often they were campaigned in more than one of these race series. Because of the confusing number of names racing Cobras have been called, we'll divide all competition Cobras into three general categories: factory team cars, factory prepared competition cars and independently prepared race cars. Cars in the first two categories can be further subdivided by the race series for which they were originally intended.

FACTORY TEAM CARS — full-specification competition models prepared by Shelby American and either raced as "work cars" or as independent but factory sponsored cars (i.e. Bob Johnson, Tom Payne, Dan Gerber cars of late 1964-1965).

CSX2002	
CSX2008	
CSX2009	
CSX2026	
CSX2127	Sebring car
CSX2128	Sebring car
CSX2129	Sebring car
CSX2131	LeMans car
CSX2142	LeMans car
CSX2136	LeMans Replica
CSX2137	LeMans Replica
CSX2138	LeMans Replica
CSX2166	427 Prototype
CSX2189	
CSX2196	prototype 427 "Flip-Top"
CSX2259	FIA roadster
CSX2260	FIA roadster
CSX2301	FIA roadster
CSX2323	FIA roadster
CSX2345	FIA roadster
CSX2430	
CSX2431	USRRC roadster
CSX2488	USRRC roadster
CSX2494	USRRC roadster
CSX2513	USRRC roadster
CSX2557	USRRC roadster
CSX2558	USRRC roadster
CSX2286	Daytona Coupe
CSX2287	Daytona Coupe
CSX2299	Daytona Coupe
CSX2300	Daytona Coupe
CSX2601	Daytona Coupe
CSX2602	Daytona Coupe

FACTORY-PREPARED COMPETITION CARS — full-specification competition models prepared by Shelby American and sold to individuals; may have occasionally run with limited factory sponsorship. (List includes cars known to date and documented; more could exist.)

CSX2017	Riverside Replica
CSX2130	LeMans prototype
CSX2154	LeMans Replica
CSX2155	LeMans Replica
CSX2156	LeMans Replica
CSX2290	
CSX2472	
CSX2385	USRRC roadster*
CSX2409	USRRC roadster*
CSX2458	USRRC roadster*
CSX2459	USRRC roadster*
CSX2514	USRRC roadster*

**although referred to here as "USRRC roadsters," these cars were actually built primarily for SCCA A/P racing.*

INDEPENDENTLY-PREPARED RACE CARS — essentially street models converted into competition cars (may or may not have been built to then-current factory specifications); converted either by the original dealer, private race shop or team, or by an individual; may have been converted when brand new or may have been converted by a subsequent owner at any time throughout its history. (Some cars in this category may have been factory-prepared; however, documentation of such has not been presented.)

CSX2010
CSX2012
CSX2015
CSX2022
CSX2027
CSX2029
CSX2051
CSX2077
CSX2153
CSX2157
CSX2162
CSX2173
CSX2200
CSX2218
CSX2237
CSX2238
CSX2240
CSX2264
CSX2294
CSX2363
CSX2378
CSX2407
CSX2448
CSX2472
CSX2473
CSX2476
CSX2566
CSX2577

CSX2142, entered by Ed Hugus and driven by Hugus and Peter Jopp. It retired after ten hours at LeMans in 1963. When Shelby returned the following year with a pair of Coupes, he would create a greater stir.

The first Cobra assault on the famed LeMans 24-Hour Endurance Race was in the summer of 1963. Shelby had AC Cars build two 'LeMans Cobras' which competed in that race. The first LeMans Cobra was entered under the AC Cars' banner and ultimately finished a respectable seventh, behind six Ferraris. The second was entered by a Shelby-franchised independent dealer-racer from Pittsburgh, Ed Hugus. This car was running 13th in the tenth hour when a connecting rod decided to vacate the premises. Six 'Le Mans Replicas' were also produced. The first series of three consecutive cars were used by the Shelby-American factory team in the 1963 USRRC series. The second three cars were sold, in typical Shelby-American style, to private parties and were campaigned in the SCCA A/Production races.

The LeMans Replicas represented a transitional phase, somewhere between the early competition Cobras that appeared at Sebring in 1963, and the cars that eventually evolved as 'FIA' roadsters in 1964. The three LeMans Replica team cars appeared in Goodyear tire advertisements and their shortened trunks were a curiosity. It was one of these cars that became the first American car to win an FIA international race.

The two LeMans Cobras were fitted with special aluminum hardtops to improve their aerodynamics and thereby increase top speed. It might even be said that they were the forerunner of the Daytona Coupes. The use of these tops — longer than the street tops — necessitated the

AC Cars, Ltd.'s 1963 LeMans entry was CS2131, driven by Peter Bolton and Ninian Sanderson. It finished 7th overall and 3rd in the GT class. Aluminum hardtop contributed to car's success, allowing it to reach a higher top speed on the long Mulsanne Straight.

One of the six LeMans Replica competition roadsters, CSX2136, is shown here at the Riverside Grand Prix in October, 1963. Driven by Lew Spencer, it finished 3rd. Subsequently, it was sold to Ed Leslie who campaigned it in SCCA events, winning the 1964 A/Production National Championship.

fabrication of special shortened trunk lids that were hinged at the bottom and swung down when opened. The cars were rarely run with these hardtops outside of LeMans, so their most distinguishing feature was the shortened trunk. Other subtle differences include larger side louvers and vents (all subsequent Cobras — street and race — were given side vents shortly after the LeMans car and LeMans Replicas), Dunlop alloy wheels, larger brake calipers, a long hood scoop that ended at the rear of the hood and extra sheet aluminum 'pants' at the rear of the front wheel wells and at the front of the rear wheel wells (to conform to FIA fender regulations). The AC car was right-hand drive, and although entered and campaigned by AC, it was originally built for Shelby (although not for export). It carried a special serial number that had the prefix 'CS' instead of CSX' (the 'X' deleted because the car was not intended for export). The Hugus car, CSX2142, was left-hand drive.

The hardtops, semi-fastbacks with larger rear windows, were removable and had a quick release fuel cap located on the right rear of the top, just forward of the top of the rear window. It was connected to a larger fuel tank by a 4-inch diameter neck. The tops also had vent slots on each side to aid cockpit cooling by removing inside air as well as equalizing inside pressure and thereby preventing the rear window from popping out.

All of the LeMans Coupes and Replicas used many components that were later incorporated into the 1964 FIA roadsters, such as front and rear anti-roll bars and revised steering arms. All LeMans Replicas were delivered to Shelby American in primer with black interior trim. Each had Halibrand pin-drive magnesium wheels (the USRRC cars used the slotted 6½" wide front/7½" wide rear wheels while the SCCA cars got the six-spoke 6½"/7½" wheels), front and rear brake scoops, oil cooler, rear end cooler, competition radiator with shrouding, Sun Electrical tachometer, flared fenders, 'spats' (flared bulges behind the front wheels), fully upholstered interior, fire extinguisher, electric fuel pump, drive shaft loop, competition roll bar (offered in either black or chrome), 37.8 gallon fuel tank, full competition engine with 48mm IDM Weber carburetors, Ford valve covers, Derrington 'Nassau' headers, and a special extra capacity oil pan. Not all LeMans Replicas were identical, but some shared interior details such as toggle switch panels, seats and fuel pressure gauges. The only LeMans Replica to compete with a LeMans hardtop was Dave MacDonald in CSX2138 at the USRRC at the Continental Divide Raceway. Other differences between USRRC cars and the SCCA cars were dependent on the regulations of each series: SCCA cars ran with grilles and without full windshields — USRRC cars ran just the opposite. The LeMans Cobras and Replicas were the first actual competition cars built by AC. Prior to this, all Cobras left AC Cars the same — in street trim. Factory built team cars or customer-ordered competition cars were in effect, street cars that were converted to competition specifications at Shelby-American. Beginning with the LeMans cars, AC made a distinction between street and competition cars at the beginning of construction.

After sorting out the Cobras during the 1963 season and tasting the sweetness of an FIA victory, Shelby decided to launch a full-scale attack on the FIA's GT class. Along with the slick Daytona coupes, a team of five special 'FIA Roadsters' were constructed to comply with FIA GT rules. The first two cars were received at the Venice plant in January of 1964, Just prior to the first race of the season — the Daytona Continental. They were all painted Viking blue and each had a different colored stripe painted across the front fenders and over the nose — to help the crews differentiate between the cars. (1964 stripe colors were red, white and yellow, with maroon being added later; a fifth car was built after one of the originals was destroyed during the season.)

The FIA roadsters can be differentiated from all other competition cars by the 'FIA dimples' on the trunk. FIA Grand Touring rules stated that each car had to be able to carry a suitcase in its trunk. At technical inspection, prior

CSX2345 was the only FIA roadster campaigned in both 1964 and 1965 by the Shelby American team. Shown here, it is driven by Phil Hill in the 1964 Targa Florio.

to each race, the 'FIA suitcase' was placed in the trunk of each car. In order for this suitcase to fit in the Cobra's trunk, two dimples, or bumps, were formed in the trunk lid. No other competition roadsters had this feature. FIA roadsters also had 'cutback' doors, the rear edge of which appeared concave rather than convex as on the street cars. The reason for this feature was the extensive rear fender flaring with which normal doors could not be used. The front fenders were also prominently flared and the cars had large oil cooler scoops cut out under the grille openings.

The FIA roadsters ran the six-spoke pin-drive Halibrand magnesium competition wheels and occasionally had grille guards that were little more than tubing bent in an oval shape. All cars had quick-jacking pads instead of bumperettes and special dashboard and instrument layouts. The Sun tachometer was directly in front of the driver, flanked by a large diameter SW oil pressure gauge on the left and an oil temperature, ammeter, water temperature and fuel pressure gauges on the right. A row of toggle switches ran across the bottom. There was no glove compartment, and a

CSX2259 was one of five Cobra FIA competition roadsters that competed for FIA points during the 1964 season. Only one roadster was used in 1965. All of the 1964 cars were painted Viking blue, a light blue metallic; single stripes across their front fenders and noses permitted mechanics and scorers to distinguish between the cars. Bob Johnson is shown at Daytona.

plaque identifying the car as a Shelby American Cobra was installed.

One other option included on the competition cars was larger brakes. They were then employed on all FIA and USRRC cars, except cars used for SCCA racing, where these larger brakes were not legal. These brakes were also used on the later 427 competition and S/C cars.

Although it looks nothing like the Cobra roadster, the Daytona Coupe is probably the most famous Cobra of all. When it debuted at its first race, the Daytona 12-Hour on February 16, 1964, it sent quiet shudders through the Ferrari team. They immediately began clucking, "It won't last" and "Cobras don't handle." But such comments, made with smirks and puffed chests did little to hide their apprehension. Shelby had upped the ante and it was a new game. Indeed. By the two-thirds mark, the lone Viking blue #10 Daytona Coupe had stretched out a considerable lead on the rest of the field. Only a freak fire during a refueling stop ended the coupe's charge and dashed Shelby's hopes for a victory on its first outing. Dan Gurney, in an FIA Roadster, CSX2259, had placed fourth in GT class behind a string of Ferrari GT cars, but the real damage to Ferrari had been done by the coupe. Enzo's invincibility had been shattered and all eyes were on the Cobras. Especially that smooth, graceful coupe.

The story behind the Daytona Coupes is about as interesting and colorful as the racing history that they wrote. Shelby began heading down the road towards the coupe in early 1963 at LeMans. The semi-fastback LeMans Cobras proved to be faster than the open roadsters. The Ferrari GTO's and Aston Martins were coupes. Shelby needed a Cobra coupe and FIA rules left the actual body configuration in the GT class open as long as the original chassis and engine were unchanged. The resident designer at Shelby-American was first-employee and jack-of-all-trades, Pete Brock. Brock sat down with a clean sheet of paper and the basic Cobra chassis dimensions and came up with an aerodynamic coupe body. When Shelby's amassed talent saw what young Brock's conception of a Cobra coupe was, they were far from impressed. Shelby, himself, was even moved to seek an outside opinion. A leading southern California aerodynamics expert took one look at Brock's drawings and said that in order for that car to go even 160 mph, at least 450 horsepower would be required. The coupe was designed around the 380 horsepower competition engine. After the expert left, Shelby asked Brock — who had remained silent through the whole visit — what he thought. Brock said, in effect, "Trust me." Shelby did; actually, it was too late to scrap the design and start over. Brock and the coupe project got the green light but when Pete turned around to gather up all of the fabricators, welders and sheetmetal men that he would need, they were nowhere to be seen. All were 'busy' on other things that just could not wait. None, it seemed, wanted to become associated with a project that was destined for certain failure. So the first coupe was built by a second-string team. Chief engineer Phil Remington and Competition Director Ken Miles pitched in and assisted wherever they were needed because it was, after all, a team effort — and something would be needed against those red Ferraris. Miles actually designed the Coupe's modified chassis.

The Daytona Coupe was born in the Twilight Zone of the then just emerging concept of aerodynamics—using the car's body shape to cut through the air more effectively and

The first Cobra Daytona Coupe, CSX2287, was driven at Daytona in 1964 by Bob Holbert (shown) and Dave MacDonald. No car produced or developed by Carroll Shelby has acquired more mystique than the Coupe. Its lines are pleasing to the eye and it won the World Manufacturers Championship—what more could you ask for? Exclusivity? Consider they only made six.

efficiently, without creating a lot of power-robbing wind resistance and drag. It was like getting free horsepower. An aerodynamically 'right' car could go faster, handle better and use less fuel, to boot. But the science of aerodynamics, as it applied to racing cars, was still in the trial-and-error stages. Theories and books had not yet been written. Pete Brock and the Daytona Coupe helped to write them.

Beginning with a bare roadster chassis, which was shipped from AC to Venice, the frame was strengthened and triangulated at all critical points. A plywood 'buck' was constructed on another chassis and it was over this that the body panels were formed. Starting with a clean sheet of paper, Brock gave the coupe a lower center of gravity, reduced frontal area and a lower silhouette by tilting the radiator forward. Recessed driving lights flanked the grille opening which let air into the radiator and a depression behind it, just forward of the engine, let the air escape back into the airflow. The windshield was raked back and a long, sloping fastback ended in a sliced-off Kamm-back tail. Headlights were covered with clear plastic which conformed to the fender lines. The entire front hood and fenders, or bonnet, as it was called, was one piece and was hinged to tilt forward for easy access to the engine and front suspension. The exhaust was routed through special headers that joined in twos and exited through four 'mufflers,' one pair on each side. They were recessed into the rocker panels under the doors. A full-length belly pan ran under the chassis.

The inside was all function — as befits a race car. It was painted flat black to resist glare. Necessary instruments and switches were grouped together on the panel in front of the driver, except for the speedometer which was relegated to a position in front of the passenger seat. The spare tire was mounted behind the seats on the floor, under the large rear window. It was hinged at the top and it opened, hatchback-style. A full roll cage protected the driver and strengthened the chassis. The 380-horse Cobra full-competition engine, with high-rev camshaft and 48mm IDA Weber carburetors powered the Coupe. Its power-with-reliability is a key factor that was designed in. Sliding plexiglass side windows were used and to the rear of them, clear plexiglass scoops ducted air in and down to cool the rear brakes. An oil cooler was mounted in the rear of the car and air was drawn into it from behind.

When the Daytona Coupe made its initial appearance at Daytona in 1964, it was painted Viking blue—a light blue metallic. By LeMans, in mid-summer, the team Coupes had two slim white stripes down the center. For the 1965 season, team cars were painted darker Guardsman blue, with the now-familiar twin wide white stripes running over the center of the entire car. Wheel centers and spokes were also painted white for matching contrast.

There were two other Daytona-type Cobra Coupes built. AC Cars built an 'AC LeMans Coupe' for the 1964 event. It was based on a 289 chassis and followed the general shape of the Daytonas, but was slightly longer and heavier-looking. It had 'eyebrows' over each wheel opening and was painted green. An accident forced the car out early in the race. A second Cobra coupe was constructed by the John Willment team, famous in England for campaigning Ford Galaxies and Cobras. Painted in the Willment team colors — red with narrow white stripes — this car was built sometime after 1967. It was very similar in appearance to the Daytonas, except that it had a lower roofline and more angled windshield.

All told, six Coupes were constructed. All six never raced at the same time, as the last two were completed for the 1965

Shelby, Miles and Brock and the beginnings of the Daytona Coupe. Chassis belongs to the competition car that Skip Hudson wrecked at Daytona in February of 1963 (serial number unknown). Brock worked from line drawings and dimensions such as he could take because there wasn't time to draw up a complete set of blueprints.

Brock at Cal Metal. Roofline, at this point, was what Brock had in mind. However, as construction progressed, it changed a little; it became slightly higher and a little more contoured.

At Riverside, the Coupe made everyone smile as it exceeded everyone's expectations. Cobra at right is CSX2166, a 289 roadster fitted with a 427 engine. It was driven by Ken Miles at Sebring.

season. The very first coupe, CSX2287, was built entirely in southern California at the Shelby-American Venice plant and at Cal Metal Shapings. The remaining five coupes were constructed in Italy. The chassis were completed at AC Cars and shipped to Venice, where they received modifications similar to those on the first Coupe. From Venice, they were shipped to Carrozario Gran Sport in Modena, Italy, where the bodies were fabricated and installed. Pete Brock supervised the construction of three coupes: CSX2287, CSX2601 and CSX2602. Shelby-American's chief engineer, Phil Remington, supervised the building of the other three: CSX2286, CSX2299 and CSX2300. The second coupe built — and the first one built in Italy, CSX2286, differed slightly from the other Coupes because the Italian coachbuilders who constructed the car inadvertently placed the cowl hoop tube 2½-inches too high. The effect was a raised windshield line and a slightly different roofline. Later this was remedied and all subsequent Coupes were constructed correctly.

Body buck was made out of plywood. Once it was completed, it was taken to Cal Metal Shaping where the actual skin was constructed. While this was going on, chassis # CSX2287 was being modified to accept the new skin. It was also being significantly strengthened.

Finished Coupe is readied for testing at Riverside. Piece in center of radiator exit duct was later eliminated because it restricted air flow.

The second race for the Coupe, Sebring, ended in victory. Even so, it was clear the car needed some more development. It was so hot that drivers almost passed out; vents were quickly added.

After the first Daytona Coupe had proven itself, plans were made to produce a few more. AC built three chassis that were shipped to Shelby American where the same modifications that CSX2287 received were duplicated. Then, the first Coupe and the three modified chassis were shipped to Modena, Italy to Carrozario GranSport. By LeMans, in June of 1964, the second Coupe, CSX2299, was ready. The third Coupe, CSX2286, was completed for Goodwood in August. The fourth, fifth and sixth cars were built in the off-season and were all completed by February, 1965, for Daytona.

AC Cars, Ltd. built its own Coupe for LeMans in 1964. It was based on Shelby's Daytona but featured deviations in some body lines. It was involved in an accident during the race and failed to finish.

The Coupes were pressed into service as they were completed. CSX2299 first raced at LeMans in 1964. After that, other coupes were entered as they were readied.

On November 6, 1965, after Shelby had won the championship and the Coupes were withdrawn from FIA competition, land speed record king Craig Breedlove took a lone Cobra Daytona Coupe, CSX2287, to the Bonneville Salt

Bonneville Salt Flats, Utah: November 6, 1965. Craig Breedlove sets 23 national and international records in CSX2287. In 12 hours, the Coupe covered 1,800 miles at more than 150 mph (average). It used pump gas from a local station and less than a quart of oil was consumed!

Flats. Running the ten-mile Bonneville Oval, Breedlove and co-driver Bobby Tatroe set 23 new national and international records. During the 12 hours of running, the drivers covered 1,800 miles and averaged a hair over 150 mph. The Coupe ran on pump gas from a nearby gas station and used less than one quart of oil!

The Cobra Daytona Coupes were built and raced in a fishbowl. The world — looking through the pages of every automotive publication — saw the cars' ordeals and triumphs. Nothing could be hidden. Shelby ran an open shop where journalists were concerned, and they recorded the Coupe's progress as it was being built and raced. When the first Coupe caught fire at Daytona in 1964, the world saw it. And when the same car crossed the finish line first, a month later at Sebring, the world saw that too. There was very little mystery surrounding these cars. They were straightforward in concept and design. And they worked.

The eleven USRRC Competition Roadsters were among the last competition 289s built. They represented the final state-of-the-art in competition Cobras and were used primarily as Shelby-American team cars in races in the United States in 1964 and 1965, notably in the United States Road Racing Championship series, although five cars were sold, brand new to customers who raced them in SCCA events.

The 'USRRC cars,' as they have come to be known, incorporated many FIA Cobra roadster features; most notably, the larger rear fender flares and the cutback doors. They used the six-spoke pin-drive Halibrand competition magnesium wheels and had standard Cobra dashboard layouts with larger competition gauges. They also carried an oil cooler scoop under the grille opening and had hood scoops with sculptured leading edges. A large competition snap-open fuel filler cap was located to the right of the rear cowl, behind the passenger seat. Roll bars, competition brakes, quick-jacking pads, larger capacity fuel tanks and most of the other competition features that were employed on the LeMans Replicas and the FIA Roadsters were also used. After their use as team cars during the 1965 season, they were subsequently sold as 'used race cars.'

Shelby-American came to within a hair's breadth of winning the FIA World Manufacturers Championship for GT cars in 1964. They succeeded in winning the championship in 1965. Cobras raced in a total of 27 FIA races between 1963 and 1965, and they set either fastest GT qualifying lap or fastest GT race lap in 26 of those races. In the 27th, LeMans in 1963, a Cobra set the over-3 liter GT race distance record. Cobras scored a total of 17 overall GT class wins and established 42 international records. And they made Shelby's dream come true — they blew Ferrari's ass off!

Before sending one of the FIA roadsters back to the U.S. following the 1964 season, it participated in the British Drag Festival. Shown here racing against a USRRC Cobra prior to its being shipped to the U.S., the fast snakes wowed 'em. Imagine what a Dragonsnake would have done . . .

SPECIFICATIONS OF THE SHELBY FORD COBRA (1964 model)

Dimensions
Wheelbase: 90″
Overall length: 167″
Width: 61″
Height to top of windshield: 45″
Tread, front: 53.25″
Tread, rear: 52.75″
Ground clearance: 5.5″

Engine Specifications
Type: High-performance Ford Fairlane V8
Displacement: 289 cu in
Bore and stroke: 4.00 x 2.87″
Compression ratio: 10 to 1
Basic carburetion — Single 4v
Bhp: 271 @ 5,800 rpm*
Torque: 269 lb. @ 4,800 rpm
**Optional versions up to 370 hp*

General
Curb weight: 2,100 lbs.
Weight distribution: 49/51
Steering wheel turns lock to lock: 3
Turning circle: 33′3″
Tire size, front: 7.35 x 15
Tire size, rear: 7.35 x 15
Brake lining, swept area: 550 sq in

Transmission: 4 speed, close ratio
First: 2.36-1
Second: 1.61-1
Third: 1.20-1
Fourth: 1-1

Differential
4.56: 110 mph
4.26: 115 mph
3.77: 132 mph
3.54: 140 mph
3.31: 150 mph
2.72: 180 mph

Top speed at 6,500 rpm with stock equipment

Features
All aluminum hand-formed body; genuine leather upholstery; individual bucket seats, limited slip rearend; rack and pinion steering; 72-spoke wire wheels with knock-off hubs; full independent suspension, front and rear; alternator, replacing generator; wide based wheels for better handling; LeMans-type 12-inch Girling disc brakes on all four wheels with dual master cylinders; choice of final drive ratios from 2.72:1 through 4.56:1. Road equipment: safety glass windshield; folding top and side curtains; turn indicators; full lighting
Colors available: red, maroon, white, black, bright blue, princess blue, silver

Price
P.O.E. Los Angeles, $5,995.00

FACTORY-INSTALLED OPTIONAL EQUIPMENT AND ACCESSORIES

(As of January 8, 1964. Prices and specifications subject to change without notice)

Group A
Tuned air cleaner, chrome: $4.00
Aluminum rocker arm covers: $52.35
Front grille guard, chrome: $39.50
Rear bumper guard, chrome: $42.50
Exhaust pipe tips, chrome: $5.00
Adjustable wind wings: $22.50
Tinted sun visors: $19.50
Smith heater: $95.00
Competition seat belts (each) $16.50
White sidewall tires: $42.00
Luggage rack, chrome: $42.50

Group B
Wheels, chrome (5) $150.00
Aluminum 4v intake manifold: $71.00
Outside rear view mirror: $6.00
Dual 4v carburetor and manifold: $243.00
Custom AM radio and antenna: $58.50
Competition oil pan: Available soon
Oil cooler: $128.25

Competition vehicles and related high performance equipment (prices available on request)

COBRA 289
Competition version

Tread, front: 52″
rear: 53.5″
Wheelbase: 90″
Overall length: 151.5″
Overall height: 49″
Overall width: 67″
Weight: 1,912 pounds
Construction: Large diameter steel tube frame of simple ladder construction, body of aluminum panels over light gauge steel tube formers.

ENGINE: Ford
Type: V8 high performance; capacity, 289 CID
Firing order: 1,5,4,2,6,3,7,8
Bore: 4.00″
Stroke: 2.87″
Cylinder block: cast iron
Cylinder heads: cast iron
Crankshaft: Cast iron, supported in five main bearings with external counterbalances
Bearings: Copper-lead
Main bearing diameter: 2.25″
Rod bearing diameter: 2.125″
Valve operation: push rod, from cam in block
Valve diameter: Inlet, 1.88″; exhaust, 1.63″
Valve springs: 1, and 1 damper spring per valve
Valve lift: .490″
Carburetion: 4 x 2v Weber 48 IDA
Compression ratio: 11.5:1
Ignition, coil: Make, Ford Voltage, 12
Exhaust system: Nassau type tuned header system with large diameter mufflers and tail pipes

GEAR BOX: Borg Warner T10
Case material: aluminum
Ratios: High, 1:1; 3rd, 1.20:1; 2nd, 1.62:1; 1st, 2.36:1

CLUTCH: Ford, disc diameter 10.5 inches

REAR AXLE CENTER SECTION:
Salisbury hypoid mounted to chassis frame by three large diameter rubber sleeve type mounts. Open half shafts, universally joined at each end transmit power to wheel spindles. Limited slip differential standard.
Ratio, standard, 3.77:1; optional, 3.07, 3.31, 3.54, 4.09, 4.27

SUSPENSION:
Anti-sway bars fitted front and rear
Koni shock absorbers standard equipment

BRAKES:
Girling disc brakes, front and rear
Front, 11-5/8″ diameter; rear, 10-3/4″ diameter

WHEELS AND TIRES
Front 6.70 x 15 stock car specials on 6½″ rim x 15″ magnesium center lock pin drive wheels
Rear, 8.20 x 15 stock car specials on 8½″ rim and 15″ magnesium center lock pin drive wheels

CAPACITIES:
Radiator: 12 quarts
Engine: 8 quarts
Fuel tank: 18 gallons
Oil cooler, large capacity steel oil pan, rear axle cooler and circulating pump, special racing bucket seats, roll over protection hoop, seat belts, shoulder harness, special instruments are standard equipment on competition cars.

OPTIONS:
Fuel tank: 37 gallon

COBRA COMPETITION CARS

STAGE I

A dual purpose machine for the man who wants something a little bit better than stock, suitable for slaloms, autocrosses and mild club racing events.

Fitted with roll bar, anti-sway bars, front and rear, seat belts, otherwise as street specifications with standard 289 CID, 271 horsepower engine $6,275.00

STAGE II

For the man who takes his racing seriously. As above, but fitted with center-lock magnesium wide-rim wheels with Goodyear racing tires, fenders modified to accept wide tires, special steering arms, hood air scoop. With standard 289 CID, 271 horsepower engine $7,220.00

STAGE III

A team car replica. Identical in every way to the famous Shelby-American team cars which won the Manufacturers Championship, the SCCA Class A Championship, the Drivers Championship. Each car is tested personally by Ken Miles at Riverside and guaranteed to equal the best lap times established by a factory team car (the GT lap record). Complies with SCCA Production Class A requirements and homologated with the FIA as a GT car, Class III.

Fitted with roll bar, sway bars, front and rear, wide-base magnesium center-lock wheels on special hubs, strengthened chassis, Koni shock absorbers, special steering arms, heavy-duty brake pads, engine oil cooler, rear axle cooler, hood air scoop, brake air scoops, racing windshield, racing seats, quick lift jack pads, long-range fuel tanks, supplementary electric fuel pump, special instruments, special brake fluid, Goodyear racing tires, choice of axle ratio. Available only with option IV, full race engine $9,500.00

Either Stage I or Stage II cars may be fitted with any of the following options, the price of which is in addition to the base cost of the car.

OPTION II-R

Standard engine but equipped with dual quadruple choke carburetors and aluminum large capacity pan, polished valve covers $254.75

OPTION III-R

Full racing engine, magnafluxed, balanced, crankshaft reworked, pistons modified, heads ported and polished and fitted with larger valves, special camshaft, degreed crankshaft damper, reworked distributor, large capacity steel racing oil pan, two four-throat carburetors on an aluminum manifold $1,907.71

OPTION IV-R

As above but fitted with four two-barrel Weber carburetors and associated linkage $2,905.42

COBRA ACCELERATION PACKAGE - DRAGONSNAKE
CONFIDENTIAL DEALER NET PRICE LIST

September 1, 1964

DRAGONSNAKE STAGE II — NHRA

BASIC CAR: *Stage II Engine: NHRA*

300 hp 289 CID; competition cylinder heads; induction assembly (2-4v); balanced lower end; modified distributor; dress up kit; dynamometer tested

Magnesium wheels, set of 5, 3 ea. 6x15", 2 ea. 7x15"; dual exhaust headers; competition front and rear springs; hood scoop; scattershield; competition pressure plate; competition clutch disc; 4.89:1 differential gear ratio; 2 50/50 lock-down, 2 90/10 lock-up shock absorbers; seat belt; competition tachometer (4000-8000 rpm); 4-speed fully synchronized transmission; roll bar; top; tonneau cover; side curtains; full tool kit; two Goodyear tires in tubeless, 106-246-430; two Goodyear slicks, 7.10 x 15 x 8" wide $6,495.00 $7,495.00

Accessories	Dealer Cost	Suggested Retail Price
Drag Exhaust Headers	$125.90	$157.20
Magnesium wheel, 6x15"	89.50	99.50
Magnesium wheel, 7x15"	90.50	104.50
Magnesium wheel, 8½x15"	96.50	114.50
Magnesium wheel (set of 5) 3, 6x15"; 2, 7x15"	439.50	489.50
Competition rear spring re-arch customers)	14.50	23.50
Competition front spring (re-arch customers)	18.50	27.50
Hood scoop	31.60	39.50
Scattershield	98.04	106.75
Competition pressure plate	17.50	25.00
Competition clutch disc	20.26	28.95
Roll bar	89.00	107.00
Roll bar (chrome)	104.92	122.92
2:50x50 lock-down shock absorber	16.90	24.15
2:90x10 lock-up shock absorber	16.90	24.15
Tire: Goodyear all weather in tubeless 105-264-430	24.55	24.55
Tire; Goodyear slick 7.10x15x8	48.00	48.00
Seatbelt	13.00	16.50
Chrome exhaust tips (each)	1.75	2.50
Competition tachometer	93.50	110.00
English hard top	271.60	339.50
NHRA competition cylinder heads (pr.)	325.50	375.50
Camshaft (AHRA) 300R (machined)	94.80	118.70

Engines, complete

Engine 289 CID high performance V8 Stage I-D, 271 horsepower	731.34	772.64
Engine 289 CID, special V8 Stage II-D, 300 horsepower	1,595.00	1,595.00
Engine 289 CID, special V8 Stage III-D, 325 horsepower	2,595.00	2,595.00
Engine 289 CID, hi-winder special V8 Stage IV-D, 380 horsepower (AHRA only)	3,295.00	3,295.00

Drag Vehicle Cost

Stage I-D, complete, 271 horsepower	5,795.00	6,795.00
Stage II-D, Complete, 300 horsepower	6,495.00	7,495.00
Stage III-D, Complete, 325 horsepower	7,695.00	8,695.00
Stage IV-D, Complete, 380 horsepower	7,995.00	8,995.00

427 COBRA

if they drove these in Heaven, everyone would be dying to get in

On January 27, 1965, Carroll Shelby officially raised the curtain on the new 427 Cobra roadster at a special press introduction at Riverside International Raceway. This car was soon being described as the 'ultimate' high performance car and fulfilled Shelby's desire to build the 'fastest production car in the world.' Quite an accomplishment in just four years. The past had seen other cars like it in concept, but nothing like it in intensity. It immediately made comparisons worthless and surely added brilliance to the Shelby legend.

The 427 Cobra was probably the car that Carroll Shelby would have liked to build instead of the 289. The leaf spring Cobra was merely a starting point. The 427, or coil spring Cobra, was the evolution of that idea. It represented the full potential of a lightweight, aluminum-bodied, tubular framed V-8 roadster. In fact, it was so successful that it was probably the singlemost element responsible for its own demise. With a weapon so fearsome in Shelby's arsenal, what would he have gained by producing something faster or more powerful — even if the creeping federal bureaucracy would have permitted it? Precious little, when he was his own competition. Shelby held the line with this car and rather than emasculate it, he decided to simply cease all production while the car was at its zenith.

Back in 1961, Shelby didn't have the money, resources or the influence to build a car of his own design. Necessity forced him to take an already existing engine and couple it to an already existing car. After his continued success, and a whetting of the gargantuan Ford Motor Company's appetite for performance and racing victories, Shelby was in a position to create a successor to the original Cobra. Few parameters forced him to conform to previous specifications and what emerged as the 427 Cobra roadster was the end result of two seasons of 'lessons learned' in actual competition and a quantum leap forward in racing technology. True, the body shape looked much the same, but that, too, was by choice. Virtually everything under the skin was new and improved.

The 427 Cobra was initially designed to replace the 289 leaf spring Cobra, both in domestic sales and in international competition. By the end of the 1964 racing season, Shelby-American, with their team of Cobra Daytona Coupes and FIA Competition Roadsters, had come to within inches of wrestling the World Manufacturers Championship away from Ferrari. This was not merely a title that Ferrari had won — it was the title that Ferrari literally owned. For the previous twelve seasons, Ferrari had walked away with the trophy each year. Had it not been for some fast-talking political shuffling by *Il Commentatore*, Shelby's Cobras would have ended that in 1964. However Ferrari succeeded in having the Monza Three-Hour 'Coppa-de Europa' cancelled and thereby prevented the Cobra team from getting enough FIA points to beat him.

Shelby and the rest of the Cobra team returned from Europe smarting. They were determined that the 1965 season would be theirs — regardless of what cars Ferrari chose to compete with. No top-flight racing team can remain stagnant for any period of time and Shelby-American was no exception. If the small blocks came close, a big block Cobra would win. The 427 Cobra actually had four prototypes that were separate and distinct from each other. Together they evolved into what the 427 Cobra became.

The first of these 'prototypes' appeared in March, 1964, at the Sebring 12-Hour Endurance race. It really didn't cause much of a splash — for two reasons. First, because it crashed in practice, and second, because it was totally eclipsed by the 289 Cobra Daytona Coupe — which had caused such a stir a month before at its introduction at Daytona. In doing so, the coupe weighted the scales in favor of aerodynamics and against gigantic horsepower increases. More horsepower wasn't needed — just maximum utilization of what was already there. But the red #1 Cobra at Sebring, driven by Ken Miles, was something else! Where the Weber-inspired 289s had a throaty mid-range wail, Miles' 427 prototype had a bellowing roar that literally shook the other cars as it blasted past them.

With almost a hundred more horsepower on tap than the 'regular' competition Cobras — and stuffed into the same chassis — it was no wonder that Miles had difficulty holding the car on the track. He spun in practice and managed to smack the only tree for what seemed like fifty miles. Subsequent all-night pounding failed to permit the car much more than a few laps on race day before it succumbed. But the seed had been sown, and Miles — a force at Shelby-American — was the farmer.

The first 427 Cobra prototype was actually a 289 leaf-spring Cobra with a NASCAR 427 engine stuffed between the frame rails. CSX2166 was sitting around Shelby American when a bench-racing session brought up the subject of more horsepower for the 289 Cobra. They were already getting 380 and that was about the limit if reliability was needed. But, someone pointed out, a 427 Ford could put out 480 without breaking sweat. That led Ken Miles to stuff one in the nearest 289 Cobra . . . CSX2166. The car was tested at Riverside and was fast. A decision was made to bring it to Sebring, in March of 1964. Miles shared the drive with Shelby American mechanic and welder John Morton. Holes in the nose and leading edge of the hood were required to keep the beast from overheating.

The second 427 Cobra prototype was another 289 leaf-spring Cobra. It was a radically modified Cobra whose strong suit would be its light weight: about 1600 lbs. Originally powered by a 390 engine, it received a 427 by the time it was shipped to Nassau in December of 1964 to lock horns with the Corvette Grand Sport. It was called the "Flip-Top" because the front hood/fenders and the rear trunk/fenders — both one-piece sections — were hinged to flip open. It was another Ken Miles special and was quickly dubbed "the Turd" by crewmembers. Its serial number was CSX2196.

The second 427 prototype came to be known as the 'Flip Top' Cobra. Developed in late 1964 and raced at Nassau in December of that year — again by Ken Miles — this car was actually about as close to a Cobra as a NASCAR stocker is to its showroom counterpart. At first powered by a lightweight NASCAR 390 engine and later by a 460 horsepower 427, it was little more than a stripped and lightened leaf spring Cobra frame with a thin-gauge aluminum body, sectioned into panels that hinged both forward and rearward from the cockpit exposing its vital organs. It tipped the scale at about 1,600 pounds and was intended to be Shelby's ace-in-the-hole against GM's Grand Sport Corvettes. Although it failed to finish at Nassau, it was so brutally fast while it was running that there was no question of where future Cobras were going.

During the space of almost a year that separated the running of these two cars, a lot was happening at Shelby-American. Shelby had almost taken the FIA title away from Ferrari but Ford still viewed the Cobra as a hybrid — more AC or Shelby than Ford. They wanted their banner carried by a car of their own design and manufacture and as a result the Ford GT program was born. The Cobra was viewed as an interim vehicle — one that had already reached its maximum potential. During the time that it would take to make the Ford car successful, the Cobra would do. But from Ford's perspective, it was never more than an interim vehicle. Shelby, in turn, saw little in the Ford GT — save for continuing his now profitable relationship with Ford. Ford GTs would be too expensive to produce for sale in any quantities and they were not Shelby's 'babies.' He found it more satisfying to beat them with Cobras — although they were not competing in the same class. In fact the Fort GT was in a higher class. The scale between Cobra victories and Ford GT successes began to tip back and forth constantly in 1965. In Ford's eyes, when their GTs stumbled because of mechanical development problems, they were happy to have a Cobra in the winner's circle. But when the Ford GTs began to charge home first, the Cobras suddenly weren't so important.

In October, 1964, the first of two coil spring prototypes was begun at AC Cars. The suspension specifications were compiled by Ford's resident suspension expert, Claus Arning, and his Ford computer. They took into consideration all of the problems and deficits of the leaf spring Cobras and all of the assets and increased horsepower of the big block engines. Actually, four coil spring prototypes were originally planned in late August. These were to be development cars and were referred to as 'Cobra IIs.' The original plan called for four cars to be fitted with either roadster bodies (basically the same as the current 289 competition roadsters) or coupe bodies (an updated version of the Daytona Coupe). They were referred to in Shelby-American internal communications as Chassis Number 1, Chassis Number 2, Chassis Number 3 and Chassis Number 4. The target date was originally put at around November 1 for completion of the last prototype and the beginning of formal production. The first three cars were to be roadsters: Numbers 1 and 3 having 390 cubic-inch aluminum engines and Number 2 having a cast iron 427 with an automatic transmission. The Number 4 car was intended to be a coupe, also carrying a 390.

By October, the Cobra II development program was revised by extending the target dates of all four cars by one month. The lightweight 390 engine was planned for use in these prototypes because the 427s proved too difficult to obtain at that time. The lightweight 427 that was to eventually be used was described as follows:

- side oiler cylinder block
- cylinder heads same as 1965 427 4v high performance except material to be aluminum
- water pump with aluminum housing
- crankshaft damper with aluminum hub
- intake manifold same as 1965 427 4v high performance except material to be magnesium
- fabricated exhaust manifolds
- radiator supply tank same as 1964 427 4v high performance except material to be aluminum
- total dry weight of proposed engine, including fan, flywheel and radiator supply tank: 565 pounds

Two of the 'Cobra II' prototypes were actually built. Chassis Number 1 and Chassis Number 2 were 4-inch tube, coil spring frames with widened 289 FIA Competition type bodies. The fenders of these two cars were actually cut along the top edge and widened by adding a two-inch strip of aluminum. No actual 427 body bucks existed at this time. Both cars had forward-braced roll bars (which accounts for their often being thought of as 289 Cobras). Once completed, they were given 427 Cobra serial numbers (CSX3001 and CSX3002) and subsequent 427 Cobra production began (with CSX3003). CSX3001 was reportedly sent to Ford in Detroit for further development and evaluation and later returned to Shelby American where it was updated to current 427 specifications (427 grille and roll bar dry sump system and competition features) and sold. It was extensively raced in Europe. CSX3002 was also outfitted with a dry sump system — the oil tank nestled under the passenger front fender, behind the front wheel, its tell-tale oil filler cap sprouting up through the sheetmetal. It was raced briefly by Ken Miles, in the name of 'development' and by Phil Hill.

The 'Cobra II' code name eventually became the cause of concern because once attached to the prototypes, it became difficult to shake when it was decided that the new models would be officially called *427 Cobras*. The press had received previews of the new Cobra II and for a while the name stuck. Some Shelby-American employees and even some of the initial owners persisted in calling the early 427 competition Cobras 'Cobra IIs,' and many race programs referred to these cars as such. Shelby eventually had to issue a directive to all employees and dealers instructing them to cease all reference to a Cobra II immediately. All coil spring Cobras were to be referred to as *427 Cobras*.

From the initial conception of an entirely new 427 model to the production of the first car took exactly three months! And to the credit of everyone involved with the project, not one single thing had to be changed. Not one. Shelby's plan was to homologate the coil spring 427 and campaign it during the 1965 season, and production was begun at AC Cars. To lock Shelby in even further, Corvette was planning to come out with a new 427 engine option. Because they were direct sales competitors, it benefitted Shelby to stay in the same engine displacement range. Ferrari was attempting to build the required one hundred 250 LM 3.0 liter coupes, which would have been considerably faster than the 289 Cobras — both the roadsters and the coupes. FIA officials watched Shelby and Ferrari closely and neither had the required 100 cars built by the deadline, so both were forced to campaign their previous year's cars. In Shelby's case, they were the 289 Daytona Coupes and FIA roadsters. Shelby had 51 427 Cobras by April 29th, 1965, when the FIA Inspection Team made their visit. Production of the 427 race cars was stopped and production of the 427 street cars was begun under the first coil spring contract with AC.

In the midst of this the 'fourth' 427 prototype was built. It was actually one of the original six 289 Daytona Coupes whose chassis was stretched two inches (at Shelby-American in Venice) and sent to Modena, Italy for completion of a special longer body. The extra length was to accommodate a 427 big block. It was to be a prototype car that would be used to evaluate the results of the horsepower increase from the big block and also as a test car for the bigger engines that everyone knew would be required for 1966. Pete Brock relates that soon after this 'long-nosed' coupe was completed, one of the 289 team coupes fell off the front end of a transporter and was severely damaged. The

The second 427 Cobra, CSX3002, as it sat at Shelby American after delivery from AC Cars. Since no 427 body bucks existed at that time from which to make the 427 body panels, the 289 nose was used and the front fenders were cut along the top edges and a 2-inch piece of aluminum was welded in to give them the correct width. Rear fenders were FIA competition roadster style, again with a 2-inch section. Rollbar was forward-braced.

Stretched Daytona Coupe sports bubble for 427 engine. Transporter accident created a 289 Coupe shortage, so the two-inch section was removed and the 289 engine was put back in.

'427' was shortened (because FIA rules prohibited unhomologated chassis) the two-inches and the car was campaigned as a 289 coupe.

The new 427 Cobra was far superior to the leaf spring model in every respect except, possibly, gas mileage. Any faults that the earlier Cobras had were corrected. And all of the things that made a Cobra *a Cobra* were retained. Most of the modifications were necessitated by — or related to — the new engine. This is not surprising when you consider the fact that the original Cobra was very similar to the AC Bristol, except that it had twice the horsepower. Competition 289 Cobras were stretching things some, with three times the horsepower and if demands caused by another 125 horsepower on top of that were brought to bear on that poor AC Bristol frame, the first time that someone would have really laid on the car, it would probably have rolled over and died.

At first glance, both Cobras look the same. The differences are most pronounced from the front, with the most obvious difference being the grille cavity. A constant problem with the Cobras when they were raced was that the smallish mouth limited the amount of air that could be forced in through the radiator. Also, in SCCA production racing, cars had to use their production grilles. This meant that the 289 Cobra had to compete with its egg crate grille — which further limited cooling. The 427 was designed with a huge oval hole as a mouth. It was almost twice the size of the 289's. Street cars had a single narrow grille bar across the middle; so much for those egg crates! All early 427s (to CSX 3300) also had a smaller, oblong oil cooler opening just below the mouth. All cars had two vertical, oblong openings on either side of the grille area. These ducts let cool air into the cockpit on street cars and served as cooling ducts for the front brakes on competition cars. Headlight position, hoods, windshields, trunks and driver's compartments were almost identical. The 427 Cobra was, however, some 5 inches wider than its predecessor in order to accommodate the monster engine. The aluminum used to form the body panels was also different; 427s used .050" thick skin (vs. the 289's .045").

John Olsen, Shelby American fabricator who was instrumental in the Daytona Coupe project, prepares 427 engine prior to cancellation of the "427 Daytona Coupe". Stretched car is in background.

Front and rear wheel openings had a more pronounced flare, allowing for wider and taller tires. Some of the early

427 Super Coupe as it sits at Shelby American in May of 1965, following its termination. The Ford GT superceeded it.

Perhaps the supreme ultimate *Ultimate* was begun but never completed. Pete Brock designed the weapon that he believed would have gained Shelby his second World Manufacturers Championship—the 427 Daytona Super Coupe. Based on a 427 coil spring chassis, this car looked nastier than the smooth 289 coupes. When a 500 horsepower engine was thrown into the coupe equation, everything changed. The 289 coupes had been designed on the aerodynamic theory of minimum drag — making the body as slippery as possible so as not to draw off excess horsepower. When large amounts of horsepower suddenly became available, more attention could be given to maximum downforce, which resulted in better (and faster) cornering. More horsepower also made the 289 coupes obsolete, but instead of being suceeded by the Super Coupe, their place was filled by the 427 Ford GT Mk II when Shelby was given the job of making the Ford GTs come home first. The Super Coupe languished in an unfinished state as the Ford GT program was accelerated. The car was referred to as the 'Super Slug' because it was not as visually pleasing as its predecessor, but Brock maintains that it would have been faster than the 427-equipped Ford GTs. But Shelby was wise enough to see where that road would have led him, with regards to Ford.

Designer Pete Brock fits rear window in. This car represented everything Brock had learned from the previous year's 289 Daytona Coupes.

The 427 Daytona Super Coupe languished unfinished until 1979 when it was purchased as a restoration project. However, it had to be more than restored in that it had never actually been completed. Pete Brock assisted by describing how he had intended the car to be finished and his instructions were carried out. The car was completed in 1981.

COBRA 427 (street version)
Tread: Front, 56", rear 56"
Wheelbase: 90"
Overall length: 156"
Overall height: 49"
Overall width: 68"
Weight: 2,150 pounds
Construction: Large diameter steel tube frame carrying suspension mounts, body of aluminum panels formed over lightweight steel tube formers

Engine: Ford V8
Type: Push rod operated, overhead valves, operated from camshaft in block
Cylinder block: cast iron
Cylinder heads: cast iron
Bore: 4.24"
Stroke: 3.788"
Displacement: 427 cubic inches
Crankshaft: steel forging
Main journal diameter: 2.750"
Crank pin diameter: 2.439"
Valve size: Inlet, 2.185" diameter; exhaust, 1.723" diameter
Valve lift: .490"
Valve springs: coil, 2 per valve
Bearings: copper-lead
Carburetor: 2 x 4v Holley
Ignition: coil Make: Ford Voltage: 12
Capacities:
Radiator: 20 quarts
Engine oil: 9 quarts
Fuel tank: 18 gallons

Gearbox: Ford
Ratios: High, 1:1; 3rd, 1.29:1; 2nd, 1.69:1; 2.32:1

Clutch: Ford, single plate, 11.5" diameter

Rear Axle Center Section: Salisbury
Ratio, standard: 3.54:1
Limited slip differential standard equipment

Brakes: Girling discs
Front disc diameter: 11-5/8"
Rear disc diameter: 10-3/4"

Wheels: 7½" x 15" alloy

Tires: 8.15 x 15" Goodyear Blue Dot Specials

Colors available: White, Rangoon red, Guardsman blue, Ivy green, Charcoal, Silver. Black interiors.

427 Cobra (street version)
Engine: 428 CID Special Police Interceptor engine with modifications by Shelby-American. Water cooled, cast iron block w/5 main bearings. Bore 4.13; stroke 3.98; displacement 428 cubic inches; 7014 cc; compression ratio 10.0:1; carburetion 1-4v Ford; horsepower 390 @ 5200 rpm; torque 475 ft./lbs. @ 3700 rpm; electrical system 12v battery, alternator; valve train, pushrod, O.H.V.; mechanical lifters; mileage 9-12 mpg; premium only recommended.

Chassis
Tread: Front, 56"; rear, 56"
Wheelbase: 90"
Overall length: 156"
Overall height: 49"
Construction: Large diameter steel tube frame carrying suspension mounts, body of aluminum panels formed over lightweight steel tube formers.
Overall width: 68"
Curb weight: 2529 pounds

Suspension: Independent front and rear using large coil spring. Hydraulic shock absorber units incorporated in a highly developed design which virtually eliminates dive and squat.

Brakes: Girling disc, front disc diameter, 11-5/8"; rear disc diameter, 10-3/4"

Wheels: 7½" x 15" alloy

Tires: 8.15 x 15 Special Goodyear Blue Dots

Drive Train: Clutch, 11.5" single dry plate, hydraulic release transmission; 4 speed, all synchro

Gear	Ratio	Overall	MPH/ 1,000 RPM	Maximum MPH
Rev.	2:32	7.68	10.86	69
1st	2:32	7.68	10.86	69
2nd	1:69	5.59	14.98	95
3rd	1:29	4.26	19.62	126
4th	1:00	3.31	25.12	162

Rear Axle Section: Salisbury, limited slip differential standard equipment; final drive ratio, 3.31:1

Capacities: radiator, 20 qts.; engine oil, 9 qts.; fuel tank, 18 gals.

COBRA 427 (competition version)
Tread: Front, 56"; rear 58"
Wheelbase: 90"
Overall length: 156"
Overall height: 49"
Overall width: 70.5"
Weight: 2,150 pounds
Construction — Large diameter steel tube frame carrying suspension mounts, body of aluminum panels formed over lightweight steel tube formers.

Engine: Ford V-8
Type — Push rod operated, overhead valves, operated from camshaft in block
Cylinder block: cast iron
Cylinder heads: aluminum
Bore: 4.24"
Stroke: 3.788"
Displacement: 427 cu. in
Crankshaft: steel forging
Main journal diameter: 2.750"
Crank pin diameter: 2.439"
Bearings: copper-lead
Valve size inlet: 2.085" diameter; exhaust, 1.650" diameter
Valve springs: coil, 2 per valve
Carburetor: Holley 4v
Ignition: coil Make: Ford Voltage: 12

Gearbox: Ford
Ratios High 1:1; 3rd, 1.29:1; 2nd, 1.69:1; 1st, 2.32:1 (Standard)
High 1:1; 3rd, 1.19:1; 2nd 1.54:1; 1st, 2.32:1 (Optional)

Clutch: Ford, single plate, 11.5" diameter

Rear Axle Center Section: Salisbury
Ratio: standard, 3.77:1; optional, 3.09, 3.31, 3.54, 4.09 Limited slip differential standard equipment

Brakes: Girling disc; front diameter, 11-5/8"; rear diameter, 10-3/4"

Wheels: Front, 7.50 x 15; rear, 9.50 x 15 pin drive magnesium alloy

Capacities:
Radiator: 20 qts.
Engine Oil:
Wet Sump: 13 qts.
Dry Sump: 14 qts.
Fuel Tank: 42 gallons

Sway bars front and rear; special tuned exhaust system, roll over protection; seat belts; rear axle oil cooler and pump; engine oil cooler; all standard equipment on competition model.

Optional equipment:
Special racing bucket seat; shoulder harness; quick change brake pad kit; mechanical chronometric tachometer; dry sump kit.

427 COBRA DRAGONSNAKE
The 427 Cobra Dragonsnake is equipped as follows:
- Heater
- Seat belts
- Drag headers, located outside front fenders; exhaust flow is directed through front fenders, outside and parallel to body lines.
- Koni front shocks (reworked to up-lock action)
- Koni rear shocks (50-50 downlock action)
- Two (2) Goodyear Drag Slicks
- Three (3) Goodyear Blue Dot Tires
- Rear end ratio: 4.54
- Hood air scoop
- Two (2) electric fuel pumps
- Battery has been relocated behind passenger seat
- Heavy duty half shafts equipped with AN nuts and bolts to prevent shearing and/or loosening under high torque loads
- NHRA approved scattershield
- NHRA approved roll bar

The engine in the 427 Cobra Dragonsnake has been competition prepared to exact factory specifications. Every clearance and dimension has been checked and set to minimum specification. Clearances of all bearings in the lower end of the engine have been set for high rpm.

All moving parts of the engine assembly have been balanced, including the cam followers, the push rods, the rocker arms, the valves and spring retaining washers. The cylinder heads are cast aluminum and have lightened competition valves installed. The ports and combustion chambers are not polished, to comply with NHRA Rules for Stock Sports Classes. The cam can be either advanced or retarded to customer specification. The rear and front spring assemblies are modified slightly to create better ground contact on the rear springs and better lift characteristics on the front springs. A hood scoop is also added to insure cool, efficient air intake to the carburetors.

Shelby-American will soon offer as an option, a hardtop and aluminum framed side curtains. No price has been established on this option to date.

Double vision: the Mk II 289 (nearest) is distinguished by the flattened-off wheel arches, the 427 by wider, more curvacious ones— and the discreet little '427' badge

SNAKES ALIVE!

To drive one Cobra is rare: to drive two immaculate examples, one a 4.7-litre, the other a 7-litre, in one day smacks of overkill. Mike McCarthy did it – and survived

Ho hum. You know how it is. You're quietly sitting in your corner of the office, trying to work out your expenses and blend into the background, when Mr Editor (Sir) fixes a beady eye on you and says "Cobras!" You pause for a while, pondering on the reasons for this sudden and unexpected expletive, when you realise he is talking about cars. Cobras. AC Cobras, to be exact.

You consider the possibility of asking him if you could find 100 hens' teeth instead, or if this is some sort of weird initiation test, or even tell him to do something that is anatomically impossible, when you begin to think. The first thing to do is to try to confuse him. "Which one?" you ask.

"How many are there?" says Mr Editor (Sir). "Well, there's the one with the big engine, or there's the one with the absolutely *enormous* engine". "Do both" he snarls. You've just doubled the problem, smarty pants.

This is where your contacts come into play. You've met John Atkins at assorted race meetings, and watched him conduct his metallic purple Mk II 289 very rapidly indeed, invariably carry off a trophy or two, then, in the paddock at the end of the day, take off the racing wheels and put on road wheels ready for the trip home. Mr Atkins deals in cherished and rare numbers through his firm Jamesigns, which is why his car carries the number JA11. It might be worth a phone call: it is, and we've got ourselves a 289. One down, one to go.

Then you remember Rod Leach, whose ads for his Nostalgia emporium are always worth reading, and who owns that splendid beast, a twin-turbocharged Cobra registered COB 1. You enter into a little bit of plea bargaining, and eventually he too agrees. In fact he's got two for sale, one of which we can drive, and which he'll pretty up for photography. Its number is, appropriately, 427 COB, again courtesy of Mr Atkins. You organise a day and place, but both Mr Atkins and Mr Leach make one proviso: if it rains the deal is off. Are they *that* bad in the wet? No – it just takes hours to clean them!

You begin to think about what you've been forced into. You start reading up what other people have said about them. The late Michael Sedgwick in *C&S*'s A-Z is a start: 'Cobra 289. 1962-69. Prod 560. 2-seater sports. 4727cc OHV V8. Not for the sybarite is Carroll Shelby's cross-pollinated Ace with American Ford engine. It'll take you to 100mph in 14 secs, and there are disc brakes, but even with some extra reinforcement the old 1954 chassis is working hard. Not many sold over here, so expect to pay for a good one'.

'Ultimate four-wheel motor-bike'

Then the Big One: 'Cobra 427. 1965-67. Prod 360. 2-seater sports. 6997cc OHV V8. Just about the ultimate in four-wheeled motorbikes, with new coil suspension and 400bhp: there have been street-legal versions, with as much as 490bhp. They're making near-replicas in at least four countries, which should prove something'.

As you read on, certain phrases and images begin to repeat themselves. Our own John Bolster said of an early 289 that it had 'a simply shattering performance' but warned 'Driven to the limit the Cobra is not for the beginner'. He did also point out, though, that 'It is outstandingly flexible and has perfect town manners'. *Car and Driver* noted that 'The AC Cobra is not. . . sophisticated or well integrated. . . (but) the hair-curling levels of performance. . . will give big production-car racers pensive moments'. In *Autocar's* hands the car (a 289) 'rocketed forward like a missile from a steam catapult. . . at a steady throttle the handling is virtually neutral. . . in the wet the whole picture changes and a lot of restraint is needed to prevent great snaking slides all over the road'. When *Motor* got their hands on a 427 they recorded a time to 60mph from a standstill of 4.2 secs, which to our knowledge still stands, and makes the 427 the fastest accelerating road car to date. But when Leonard Setright wrote about them in *Car* he went right over the top, his vitriol claiming that 'the AC Cobra is a thoroughly, remarkably, almost unbelievably, bad car'. Meeow!

But you're getting the drift. The Cobra, in either little or big-engined form, was a machine for the brave, a straight-line sling shot, a point-and-squirt device, agricultural in its road manners, lacking in creature comforts, but still about the ultimate in macho-mobiles. One to be approached warily, in fact, in case it spat in your eyes.

Right: everyone else seems to have created the myths, let's come down to earth a little. Cobras are exciting – if you want them to be. They are, also, tractable and undramatic – if you want them to be. They are nowhere near as bad as LJKS said, but perhaps his were poor examples. After all, with hand built cars there tends to be car-to-car variations that can make an awful lot of difference. They *do* create problems, one being that the more you drive them, and experience the exhiliration of the performance, the more they can lead you into going faster and faster until you reach the point of

1986 AC Cobra

The past crashes head-on with the present

by Rick Titus

PHOTOGRAPHY BY JIM BROWN

Nostalgia is a wonderful thing, really. It gives man the opportunity for memories of his own design. The vehicle by which the conscious mind conveniently removes what it wishes to forget, and adds what is more pleasant.

Such is the case with Carroll Shelby's original Cobras. Good cars, yes. But as great a car as our memories would have us believe? Not likely. The first Cobra was, more by accident than intent, a purpose-built machine. A two-place, double throwdown, haul-the-freight sports car. An automotive monster not unlike the work of Doctor Frankenstein.

Its creator was a smooth-talking Texas hot-rodder with an international race-driving reputation, a dream, and a propensity for stepping in sheep dip and ending up smelling like a rose. Its body was wrought from the grave of AC Cars in England, a spicy little sports car of ancient design and no future. Its heart came from the bowels of Ford Motor Company. It was simple luck the AC platform made a perfect introduction for Ford's new 260-cu-in. V-8.

All this is not as nostalgia would recall it. More the norm are images of the Cobra romping to victory in national and international racing, whipping the likes of Ferrari, Jaguar, Aston-Martin, and, of course, the Cobra's archrival, the Corvette.

And perhaps more romantic images of oceanfront drives, with the leading-lady's long, blonde hair

A well-driven Cobra could put the dust to anything

trailing over the Cobra's wind-swept rear deck as it wisps its way along Route 1. These are the pictures one calls up when fantasizing of the Cobra prowess.

These are not unreasonable images, mind you, but somehow devoid of the Cobra's other traits—its truck-like steering, its furnace-hot footwells, or the second- and third-degree burns fetched from the side exhaust pipes. Not to mention its buckboard ride or Niagara Falls roof. None of these are major setbacks to the overall image of such a legend, but serve to put all things Cobra into proper perspective.

Of the Cobra's total 1500-car production, 350 were 427s. These were to become the ultimate Cobra. The single most feared, revered, and desired automobile ever built. Just as the Corvette was learning how to beat—or at least run with—the small-block Cobras, the 427 came along and knocked the Chevy's legs out from underneath it. A well-driven—and that was the key—427 Cobra could, would, and still can, put the dust to anything. It simply redefined the word performance.

As street cars, they were a bunch "overkill" and, in fact, sank their fangs into their owners on plenty of occasions. The 427 Cobra represented all things masculine, the "Rambo" of sports cars.

The Cobra ceased production by mid-1969, and by 1975 was nearly a dirty word, as America was in the midst of an energy crunch. Perfor-

mance anything, much less a car fetching under 5 mpg, was a scourge. And what OPEC hadn't done to kill performance cars, the insurance industry had.

But now, with the term performance slipping back into the vocabulary of the average car buyer, and with the help and recognition of the Shelby American Automobile Club, the Cobra has seen a resurgence of popularity that has surpassed its original fame.

To support this newfound market demand, over 50 replica manufacturers have produced Cobra lookalikes. Offered in everything from kit form to turn-key, many of these products are grim exercises; under-engineered, overstated, and just plain hodge-podged together. But given the original Cobra's value today, which runs in the range of $45,000 for a 289 Cobra in need of restoration to $90,000 for a fully restored 427, one quickly understands why so many people are trying to produce replicas.

Several firms, using a fiberglass body, have taken the time and attention to build a reasonable replica that is usually offered as a kit, although several companies offer their cars in "turn-key" form.

Which brings us to Autokraft. The British firm is currently producing a 427 replica on the original body bucks and, working under license from Ford Motor Company, is installing Ford's small-block 302-cu-in. V-8, and is setting up a distribution network through Ford dealers.

The body is built from 16 gauge aluminum, and the chassis is an exact reproduction of the original.

Sold as the AC Mark IV, the subject of our test is as close to history repeating itself as the Cobra mass market is likely to get. Built on the original tooling and of the original design, given a few structural updates for crash worthiness and shock-absorbing bumpers, it's a near exact reproduction of the original, sans the killer 427 cu in. monster motor.

The AC Mark IV, so named because AC Cars of England, the original constructor, sold Autokraft the tooling and manufacturing rights, and because Autokraft perceives this version as the fourth in terms of refinement, has profited from its 20 years of development. Unaltered in silhouette but refined for real-world street-car use, it's the perfect blend of snake-bite performance and snake-charmer comfort. Interior dimensions are enhanced with the use of a fiberglass footwell and firewall structure. The greatest benefit is more room for the driver's feet, not to mention the extra sound deadening. The seats are shaped like the originals, but incorporate DOT-required head restraints. Carpeting, door panel fabrics, and the overall trim are upgraded from the original cars.

For all those who dreamed of buying a new Cobra, their fantasies can be fulfilled

The original side exhaust that looked great, sounded mean, but removed flesh from the bare legs of the unsuspecting is gone. Now, a throaty muffler and tailpipe system are tucked safely under the car. Autokraft has installed DOT-approved three-point seatbelt systems as well as added door guard beams and extra cockpit reinforcement.

At this writing, the name Cobra does not appear on the AC Mark IV. SVO Marketing Manager John Clinard has confirmed that "Ford intends to permit Autokraft to call the car a Cobra. We've already worked out an engine and transmission program with them, helped them with the technical logistics for smog approval, and informed our dealers of our blessings with regard to selling the car. Ford would like to see the name Cobra on the vehicle." Clinard adds, "We feel it's a well enough built product that it can only add to Ford's showroom image. We're all really excited about the car."

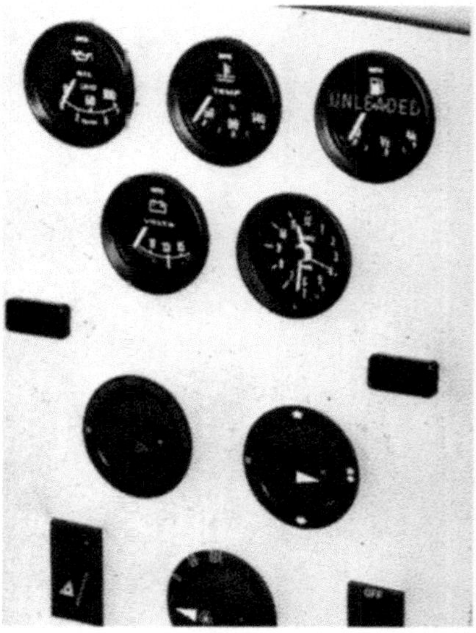

Center-lock cast wheels are a near reproduction of the originals. Note hub nut is safety wired to the wheel. Nice touch.

DOT-required side marker lights fit flush and clean. Fender-mounted engine bay exhale vents, used on the original Cobras, are very effective at lowering engine running temps.

The opposite side of the coin is that the man behind the first Cobra, Carroll Shelby, isn't really pleased about the whole program. "Hell, I've got nothing against the man's car. It's a nice piece of work. But it ain't a Cobra," he told us. "Shelby American built the Cobra and I just don't feel right about the name being on the car." Shelby speaks with as much concern for the original car owners as he does for history. He assures us his lawyers are talking plenty with Ford.

Right now, owners of the new AC Mark IV are being told to stand by, Cobra badges will be mailed to them. Most would be only too happy to adorn their hoods wth the famed snake.

As for the car itself, it's nearly perfect. The finish from body to chassis to interior appointments is fantastic.

The finish is nearly perfect, but the side curtains still leak. Some things never change

The feel is pure Cobra. The slightly enlarged interior is still small, but the seating position is adjustable and can be made comfortable for even the tallest driver. The dashboard is well laid out, without much concern for matching the original in gauge placement. The full gamut of instrumentation is there. Tach, speedo, oil pressure and temp, fuel, water, and charge. Heater and vent (no air conditioning as yet) controls are well placed in the center of the dash.

A small, leather-wrapped steering wheel comes installed on the car; however, in the trunk, carefully wrapped in plastic, is a wooden replica of the original wheel. This must be owner-installed, due to DOT's strong dislike for wooden steering wheels. Also in the trunk, which is fully finished with carpet, is a fifth center-lock wheel with a full-size mounted Goodyear gatorback GT and a tool kit.

Standard equipment for the AC Mark IV is a soft top that fits well over its lightweight aluminum bows, but leaves sufficient gaps around the side-curtains to leak enough rain to remind you from whence it came. Details of the 1980s era include a locking gas cap and electric radio antenna. Required, but nicely incorporated, are the side marker lights, flush-mounted, cleanly placed, they can easily go unnoticed.

Classic trim items like the form-shaped stainless metal rock guards on the rear fenders, the chrome T-

Inside Autokraft

At first sight it's obviously a railroad embankment, but once you've driven into the industrial center near Weybridge, southwest of London, you realize it's not a railroad but the unused banking from Brooklands. The famous, heavily banked, pear-shaped track, built in 1907, was the world's first proper auto racetrack, and was used until 1939. Today, at the edge of the crumbling banking is a ramshackle building with a sign that reads "Autokraft." There's no clue to the company's purpose in life. However, step through the door, and one is immediately transcended into the early '60s, for there is a row of gleaming new Cobras, and signs of more being constructed in the background.

But this is 1986, and AC, the English company that originally made the Cobra for Carroll Shelby, is out of the auto manufacturing business after an unsuccessful attempt to market the ME3000 mid-engine sports car. Although AC backed out of the business, it didn't stop one individual from coming to the rescue. Brian Angliss, who runs Autokraft, has been restoring Cobras for 12 years as well as building replacement parts for Cobra owners all over the world. In 1981, AC gave Autokraft the original chassis jig used to build the Cobra chassis along with the original bucks used to form the aluminum body panels. But more important, AC gave Autokraft the rights to use the AC trademark until 2006.

Instead of just reproducing the car in its original form, Autokraft took the opportunity to re-engineer and improve the car so it would meet current U.S. safety and emissions standards. By 1982, the first Autokraft AC Mark IV rolled out of the shop, and revival of the famous Anglo-British sports car had begun. Over 100 have been built and sold worldwide since then.

Ford has given Autokraft permission to sell the AC Mark IV through its dealer network, and several dealers have already agreed to stock the car. It is the first non-Ford model to be sold since the Pantera was marketed through Lincoln-Mercury dealers. Ford has given its stamp of approval to the AC Mark IV, but—perhaps more important—the Shelby Owners Club is allowing owners of the Autokraft AC Cobra Mark IV to become members, whereas replicas are not welcome. —*John Rettie*

No, it's not Shelby American at the Los Angeles Airport, 1966. It's Autokraft in England, 1986. Production is up to steam, as old world craftsman hand-form the aluminum body panels.

The AC Mark IV is as close to history repeating itself as you're likely to get

handle trunk lever, and the aerodynamically shaped chrome hood handles are there, as are the famed Cobra fender side vents for the engine compartment, and the chrome radiator splitter.

The wheels are made by Autokraft as neat reproductions of the original center-lock Hallibrands. These are fitted with Goodyear 225/60VR16 gatorbacks for the front and 255/60VR16s for the rear. Autokraft also produces its own suspension and cast uprights for the car.

A Salisbury rearend is used, as on the original, mated to like suspension componentry. Both front and rear are unequal-length wishbones with coil-over shocks. Spring and dampening rates have been refined over the original, and the ride profits from it. Suspension bushings are also uprated, as almost mandated by the use of Goodyear's gumball gatorbacks. The Mustang GT rack-and-pinion steering is used, but our guess is it's the slower-ratio setup. At 3.3 turns lock-to-lock with the smaller leather steering wheel, the car is a little slow to point. However, once induced to the stops, it has a turning circle not much larger than its length.

In terms of performance, the AC Mark IV is slightly paled compared to the original. The big, bad old 427 just flat made that Cobra nasty. But by today's standards, this stock, fully Ford Motor Company warrantied 302 pulls well. Well enough to run with any late-model Corvette you might come across. And, unlike its predecessor, it'll turn corners with the current Vette as well. Built with a predominant understeer that can be quickly overcome with throttle-induced oversteer, the AC Mark IV may be slow to point in, but takes a set and remains stable throughout the corner. Good corner departure traction permits the car to accelerate hard off a corner. Rapid transition makes the chassis twitchy, more a result of its 90-in. wheelbase than its suspension design. In an effort to make the car ride better, softer spring and anti-roll bars have slightly reduced its pure racing cornering power, but it's an intelligent compromise. →

Kid Not Spoiled Enough? Buy Him a Cobra of His Own

What's bigger than a breadbox and costs less than the air freight bill for your new AC Mark IV? Why, it's a mini-Cobra, of course. And if you step up and buy yourself a Cobra, you'd best have one in the trunk for the kids. Well, not to worry. Beverly Hills Motoring Accessories has just what you need. For $2995, you can remain at the head of your kid's "Rad Dude" list.

At just over 6 ft long and 20 in. wide, the mini-Cobra is a near reproduction of the original 289 competition car. Powered by a 5-hp 4-cycle 1-cylinder gas engine, it runs a 2.5-min quarter mile at a top speed of 15 mph. Don't laugh, it gets from 0-6 mph in less than 1 min, about 30 sec quicker than the Yugo.

With its mig-welded steel tube frame and its 13 x 5-in. tires mounted on 6-in.-wide aluminum wheels, the mini-Cobra pulls an estimated 1 g on the skidpad. Designed to hold one adult and one child, it has all the same steering, throttle, and brake controls of a regular car. Good news, no shifting necessary, it's an automatic. And if you're into twins, you can order your mini-Cobra to match your AC Mark IV, as long as your AC Mark IV is red, white, blue, black, or British Racing Green. *—R.T.*

Testing in the big chill—cold, rain and even a leaky top, and still the AC Mark IV is fun to drive. Wet weather cornering proved stable. Mild understeer is easily controlled with throttle-induced oversteer.

Cobras, past and present. Ron McClure's restored 427SC (background) serves as a reminder of the competition heritage the earlier Cobras earned. The legend wasn't built on looks alone.

The AC Mark IV is a smooth, easy to pull away from a stoplight kind of ride. Its rather tall 3.08:1 final drive ratio (made necessary to pass CARB requirements) keeps it from being a neck snapper off the line. But once placed in the "get mean" mode, it's a fully capable performer. More a long-legged runner than a dragster, its quarter-mile time and trap speed of 94 mph doesn't exactly put it on the limp-wristed list either. And once it starts to get its lungs in 4th and 5th gears, 125 mph rolls up quickly.

An oil cooler and a hood scoop are about it on the options list. Standard are 4-wheel disc brakes, Connolly leather interior, full tonneau cover, and windshield wind wings. It seems odd to see an EPA gas mileage guide glued to the right side window of a Cobra. By the way, they rated it 16 mpg city/23 mpg highway.

The AC Mark IV has a retail list price of $47,900. *Motor Trend* tested two cars, one from AC Mark IV dealer Paul Thomas of Frontier Motors in Pawnee, Oklahoma, and the other a freshly delivered gloss black AC Mark IV from new owners Greg and Debbie Moen of Westlake Village, California, who proved to be troupers when the California skies opened up and poured on our scheduled test date.

The orginal Cobra earned its stripes. It was a hybrid of awesome potential. It was part of an automotive era that is likely never to be repeated. And it did make automotive history. Bar none, the Cobra is the single most recognized automotive silhouette in the world. For the few who had a chance to drive the original, there is no going back. But for the so many who always longed for the chance to own a Cobra, the AC Mark IV is the perfect car. Seems we're not what we used to be, either. A little more comfort, while not driving quite as fast as we're used to is okay. MT

DATA

AC Cobra

GENERAL

Vehicle mfr.	Autokraft, Ltd, Surrey, England
Vehicle importer	Frontier Motors, Inc., Pawnee, Okla.
Body type	2-pass. roadster
Drive system	Front engine, rear drive
Base price	$47,900

ENGINE

Type	OHV 90° V-8
Displacement	4950 cc (302.0 cu in.)
Bore & stroke	101.6 x 76.2 mm (4.0 x 3.0 in.)
Compression ratio	9.2:1
Induction system	4-bbl carburetor-downdraft
Valvetrain	OHC, 2 valves/cylinder
Crankshaft	Cast modular iron
Max. engine speed	6000 rpm
Max. power (SAE net)	200 hp @ 3000 rpm
Max. torque (SAE net)	285 lb-ft @ 3000 rpm
Emission control	EGR catalytic converter, NOx sensor
Recommended fuel	Regular unleaded

DRIVETRAIN

Transmission	5-sp. man.
Transmission ratios	
(1st)	3.35:1
(2nd)	1.93:1
(3rd)	1.29:1
(4th)	1.00:1
(5th)	0.68:1
Final drive ratio	3.08:1

CAPACITIES

Crankcase	3.8 L (4.0 qt)
Cooling system	6.7 L (7.1 qt)
Fuel tank	64.4 L (17.0 gal)
Luggage (trunk)	192 L (6.8 cu ft)

SUSPENSION

Front	Independent, unequal length wishbones, coil springs over shocks, anti-roll bar
Rear	Independent, unequal length wishbones, coil springs

STEERING

Type	Rack and pinion
Ratio	18.3:1
Turns, lock to lock	3.3

BRAKES

Front	Ventilated disc
Rear	Solid disc

WHEELS AND TIRES

Wheel size	16.0 x 7.0 in.front, 16.0 x 8.0 rear
Wheel type	Aluminum alloy
Tire size	225/60VR16, 255/60RR16
Tire mfr. & model	Goodyear Eagle GT
Tire construction	Steel-belted radial

DIMENSIONS

Curb weight	1437 kg (2620 lb)
Weight distribution, f/r (%)	49/51
Wheelbase	2286 mm (90.0 in.)
Overall length	4114 mm (162.0 in.)
Overall width	1727 mm (68.0 in.)
Overall height	1244 mm (49.0 in.)
Track, f/r	1422/1524 mm (56.0/60.0 in.)
Min. ground clearance	76.2 mm (3.0 in.)

CALCULATED DATA

Power-to-weight ratio	13.1 lb/hp
Top speed	155 mph (est.)

SKIDPAD

Lateral acceleration	0.86 g

FUEL ECONOMY (mpg)

EPA rating, city/hwy.	16/23

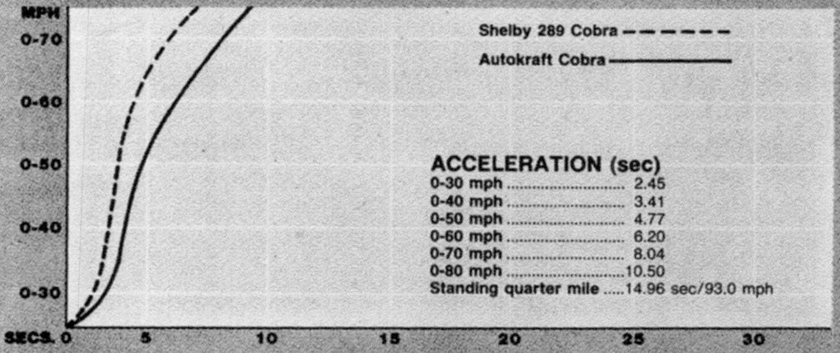

ACCELERATION (sec)

0-30 mph	2.45
0-40 mph	3.41
0-50 mph	4.77
0-60 mph	6.20
0-70 mph	8.04
0-80 mph	10.50
Standing quarter mile	14.96 sec/93.0 mph

SPEEDOMETER

Indicated	30	40	50	60
Actual	29	40	50	61

BRAKING

30-0 mph	41 ft
60-0 mph	156 ft

THE REAL THING: Phil Hill grapples with his Cobra on the Targa Florio course in 1964, the first full season for the model. He retired with a rear suspension failure, but Dan Gurney / Gerry Grant finished eighth in a sister car. Below, the "evolution" AC Mark IV is new, and built at Brooklands

Autokraft AC MkIV —

A Cobra in all but name

TWENTY years ago endurance racing was in its heyday, and the Grand Touring category was graced by such machines as the Ferrari GTO (the ultimate in classics today), lightweight E-types and Porsche 904s. Onto this scene came a newcomer, Carroll Shelby's 4.7-litre AC Cobra, even then a somewhat crude machine which proved fearsomely fast. . . so fast, in fact, that it led the Daytona 2,000 kilometre GT race overall for seven hours, driven by Dave MacDonald and Bob Holbert, then succumbing to an overheated differential. The date, Februrary 1964.

At the end of that year the Shelby Cobras were placed third overall in the World Sports Car Championship having been fourth overall at Daytona and fourth overall at Le Mans, where Dan Gurney and Bob Bondurant campaigned a coupé version, and these cars continued to give good account of themselves in the next couple of seasons, winning the GT category outright in 1965.

The Shelby Cobras were not merely scoring a lot of points, they were also marking a watershed in motor racing history, signalling out the era when drivers raced in shirt-sleeves and could be seen, easily, hard at work on the Targa Florio circuit or at Goodwood. They did not handle awfully well on fast circuits like Spa, but they were great favourites with the crowds and they made a lot of noise . . . when the big V8 engines bellowed, you could *feel* for the driver as he wrestled to keep the car pointing in the right direction!

Not surprisingly there have been a great many imitators over the past 10 years or so, replicas abounding in America, for instance, and bearing a fair resemblance to the originals. But at Brooklands, inside the boundaries of the banked track, Autokraft are in business manufacturing the real thing, genuine AC badged sports cars that have everything except the Cobra name . . . that's the property of the Ford Motor Company, and they're keeping a tight hold on it.

The Autokraft AC MkIV could not be described as a replica since Brian Angliss has negotiated with AC for the right to produce the cars, and has all the original jigs and dies to do the job; he even has some of the original craftsmen on his production line, which is turning out two cars each week. They have aluminium bodies, just as the originals did, but have been carefully modified and improved in such a way as to meet present-day legislation.

The chassis is still made up of four-inch diameter steel tubes in ladder frame style, but the dimensions have been altered to add six inches to the length of the cockpit and four inches to the width, the windscreen having moved forward three inches, so that the interior is suitable for people . . . well, Dan Gurney's size (heaven knows how he drove the original car). Each body, made of 16 gauge aluminium rather than the lighter 18 gauge of the original, takes around 200 hours to make, and the standard of trim and finish can only be described as concours. Wilton carpet and Connolly hide are used in the cockpit, leaving no doubt that this is a connoisseur's car.

To meet international construction and use regulations the bumper bars, which look very original, are mounted on telescopic struts in order to meet the 5 mph impact test. The steering column is collapsible, again to meet impact requirements, and a dual hydraulic system is used for the brakes, serving 11.63 in ventilated front discs and 10.75 in solid rear discs. The Mark IV, which Angliss describes as an evolution, "what the AC would be today, if it had remained in production", has such refinements as plastic mouldings inside the wheel arches to protect the aluminium, also lining the boot for the same reason, and a decent amount of heat and noise insulation material on the front bulkhead.

Angliss' Autokraft company has specialised for a number of years in supplying off-the-shelf parts for AC Cobras, even to the replica firms, and specialised increasingly in restoring and repairing these machines. It was only a small step, or so it seemed, for Autokraft to go into manufacturing, but the path was bumpier than it looked at first. Having negotiated the name with AC, Angliss then made a dozen cars early in 1982 and shipped these to Chicago, where the agent was willing. . . but lacked the all-important type approval.

It took protracted negotiations with Ford (helped by the fact that Walter Hayes is a friend and former neighbour of Angliss) to get the formal go-ahead, and since then Ford have been very helpful to the whole project. The American manufacturer has ordered a series of cars which will be featured in dealers' showrooms, with a price tag of $39,950 (about £30,000), and has given tangible assistance in engineering terms by developing the suspension, which is now fully adjustable with caster adjustment, Rose jointed on the top links, and stiffened.

Most of the cars that leave the Weybridge factory will be rolling chassis, leaving the power unit to the customer's choice (so long as it's a Ford, naturally). The recommended unit is the Boss Ford, the 302 cubic inch (4,984 cc) V8 which can develop anything from 200 bhp in standard trim to 400 bhp for the road, more still for circuit use. Power is transmitted by a Borg Warner T5 5-speed gearbox and a Salisbury rear axle which has a limited slip differential, and the independent suspensions at the front and rear are by unequal length wishbones and concentric coil spring/damper units.

The Mk IV we took away for a day was bound for Johannesburg immediately afterwards, the customer having specified a 350 bhp version of the Ford engine. Brian Angliss was bemoaning the fact that it didn't have as much top-end power as it should, but the brute force of the acceleration from rest left little to the imagination.

The doors, very small and light, allow restricted access to the cockpit, and the first thing you realise is that the pedals are offset to the left (in a left-hand drive model), as dictated by the enormous bellhousing and

THE Autokraft AC Mk IV has a longer, wider cockpit but faithfully copies the original lines

LEATHER trim finishes the interior to a high standard. . . but weather protection is spartan

transmission cover. The pedal box has been extended in the course of updating, which seems just as well, since you have to shoe-horn yourself in.

The steering wheel is fully adjustable for height and reach, and we chose the full up-and-out position to gain thigh clearance.

To suit all the power, Angliss fitted the strongest clutch available and this turned out to be immensely heavy, feeling like the brake pedal. The throttle was rather stiff and heavy too, and we felt that all three pedals could really have done with servo assistance!

Modernised though it may be in a few respects, the Mk IV is still a brutal car. Huge, aluminised silencers are positioned underneath, below the doors, and they only slightly alleviate the rumble of the engine which turns heads hundreds of yards away as the AC accelerates up the High Street.

The wind howls in from all directions, weather protection consisting of nothing more than a tonneau and sidescreens, and on a cold day in March there were probably many better ways of getting from Brooklands to the Chobham test track, or any other A to B journey you can think of. The first four ratios are very closely spaced indeed, fourth being ideal for pottering along at 40 mph, and fifth is a much higher gear for the open road.

On a derestricted road, or more particularly on a track, the AC takes on a completely different identity. When you get the chance to keep the throttle pressed down for more than a couple of seconds the car rockets up the road in a blare of sound, leaving twin black lines on the tarmac. . . weaving lines, from left to right and back again, interrupted by the first gearchange, illustrating how much power there is to put through the massive Goodyear Eagle tyres. For the record, they are 215 section at the front and 255 section at the rear, mounted on Hallibrand centre-lock 15 inch wheels, 7½ inches wide at the front an 9½ inches wide at the rear.

"You should have been able to get wheelspin in third gear as well", Angliss told us later. That wasn't really on, the way the car was set up, but it would have been of fairly academic interest because it would have called for unusual dexterity to keep the car pointing straight ahead at upwards of 70 mph. People like Sir John Whitmore and Richard Attwood were rather good at that, and they're welcome!

We had expected the steering to be rather more direct than it was, and this revealed that a lower geared rack is now fitted with 3¼ turns from lock to lock, rather than the original 2¾ turns. Even the new specification sounds all right, and the steering certainly isn't particularly heavy, yet our choice would still be the heavier and more direct ratio, for as the tail lashes under acceleration the driver has to work away pretty hard at the steering wheel to keep pace with events.

The brakes felt very strong indeed, though the pedal travel increased a little when they got really hot. The speedometer was disconnected so we had nothing to calibrate to bring you a set of searing acceleration figures, though an American magazine has recorded 4.3 sec to 60 mph with a 300 bhp engine, which is a full second quicker than a Porsche Turbo.

Within the limits of adhesion the AC felt safe enough on the test track, understeering a little with power applied, and to round the day off we took it onto the steering pad to see what vices could be found . . . and it was very reassuring. Understeer builds up gradually until the front wheels are sliding, and increasing the power progressively brings the tail round until the whole car is drifting in a nicely balanced fashion. You *can* apply too much power and spin the whole thing round, but it would take a very foolish driver to do this by mistake, a spin being the more likely outcome of lifting off too abruptly with a lot of cornering force applied.

Modern-day cars are a lot more sophisticated in terms of comfort, ride and handling and no amount of tinkering could make the AC anything but a period piece. The ride is very firm indeed, though there's no hint of bottoming on broken surfaces, and the whole car fits together as a superb throwback to the Sixties, the era of raw, rugged power and to hell with Nader!

If you wanted one today you'd have to join a long queue behind Americans, Germans, and a whole lot of other nationalities (there is even one being built for Japan), and as it lacks British Type Approval, getting it registered would be something for the customer to negotiate. To be honest, if we could afford a car like that, we'd ship it out to a nice warm climate and keep it there for holiday motoring. It is nice to know that one company is keeping the heritage of Brooklands alive, and helping Britain's balance of payments. — M.L.C.

SLEEK and sensual, the AC cleverly incorporates new safety features.

AC MK IV

It sure looks like a Cobra to me

STORY & PHOTOS BY JOHN LAMM

Your hair gets whipped about and tangled. Wind cuts in over the windshield and buffets you about. Sounds roar at you. All your senses are burning bright.

Nothing wrong with that.

In seeming defiance of this techno-microchip age, it's a simple beast. Just a tube frame wrapped with a carefully formed aluminum body. V-8 power. A lot of horsepower lightly burdened. Fat tires. Big brakes. A basic suspension that gets the job done.

Nothing wrong with that either.

Your foot on the throttle is like a foot on your chest. The harder you push down, the more it pushes you rearward. Hold on, because here comes 60 mph. There it goes. Here comes 100. Pure, unbridled acceleration.

And there's certainly nothing wrong with that.

YOU SHOULD SEE the place when it's empty and quiet. Like on a sunny Sunday afternoon, when the large skylights wash the rooms with warm light. Partially finished aluminum body panels rest in the metal fabrication shop where the workers left them on Friday. Some panels remain on body bucks. A space frame that still lacks a few

7803 P

Around Goodwood the GTO initially feels uncooperative, but relax with it and handling genius emerges. This car can run on the road too, hence big speedo in front of gearlever (below left)

from a lightweight 300 horsepower engine in the front driving the rear wheels of a car weighing just over 1000kg. It's a combination the Ferrari had by birth and you can even lose it if you stiffen up the front springs and make it more like a modern car, which is what some people do. You can brake later, sure, and turn in harder, but then the tail slides until eventually you have to come back off the power.

At the same time as you learn how to use the other controls, your lack of inhibition means you have relaxed the biceps and slackened the grip on the aluminium knob. Now you just aim and flick, and the gate guides the lever perfectly while the synchros offer no resistance. The shift is swift, the ratios close, the organ chord shifts effortlessly down the keyboard. Each lap you think you can make everything sing with more harmony, dance with a daintier foot. You have discovered the magic. How could the Cobra follow a performance like this…?

It starts by offering better accommodation. The seat is similarly minimal, equally cracked and leathery, but it slides back and the wheel is clear of my thighs. Against that, the gearlever poking vertically from the tunnel needs thought rather than instinct. Then the engine spins more slowly on the starter, catches with similar indifference, then bellows and rumbles, sending a thrum through the frame. Picks up mighty smartly though – the slightest touch of the pedal flicks the needle round the dial and rocks the whole car – but the real difference is the way it sends the Cobra down the road. The V8 feels half as muscular again, pulls like a glider winch from nothing and pins you to your seat whatever the gear. There's that instant and massive response only found in genuinely powerful cars, where you add a bit more boot and the car always leaps forward as if there's a giant outside force sweeping it along. We did the sums later and found the Cobra was pulling 151mph at the end of the back straight but the way the pile of railway sleepers ahead starts to fill that flat screen leaves you in no doubt of the energy involved, just as it does when you approach St Mary's. Meanwhile the sheer V8 grunt and off-beat rattle soon becomes as addictive as the Ferrari's wailing V12. The grunt gives you other options, while at the same time creating its own problems.

'The sheer V8 grunt and off-beat rattle soon becomes as addictive as the Ferrari's wailing V12'

The Cobra feels heavier at the front than the GTO and the steering is weightier, with lots of kickback. You have to wind up what feels like ten degrees of spring-loaded slack to overcome the initial push and make the nose point, then having succeeded, it will promptly unwind and push again. These are only small movements mind, much like the fidget in the Ferrari, but without the drift and I couldn't seem to get one of those going. There was never the sharpness that eventually came to the GTO's helm and never the calm that descended once you got the yaw started. But there were the other options. You don't need to take so much speed into the corner and keep the Cobra's engine on the cam, because it always is and you don't need to grapple a slower gearshift anything like as often. You can leave it in the higher ratio and use the vast well of torque to accelerate through the corner in the traditional slow-in, fast-out sense – added to which you just drive the car according to what it's telling you rather than having to apply a technique which feels alien at first. It's more physical but generally easier to access. That said, you do have to be careful.

Backing off the power suddenly in a corner had to be avoided at all times because it gave you the kind of oversteer you get from yanking on the handbrake. Then towards the end of my second stint I had found a good balance in the first part

Cars like these marked the end of an era – race-cars were becoming more specialised; soon it would be mid-engined Ferraris and Ford GT40s doing battle

of Lavant – the double-apex sequence which leads onto the back straight – and while it wasn't quite drifting, the Cobra was driving hard, just on the point of oversteer when, suddenly, the rear end did a mighty jig, almost as if something had broken. The revs soared and the tail snapped to the left and although I caught it, it was a messy flop back the other way. Worst of all I had to lift… and it was such a good lap. It was either the leaf spring on the left had run out of up travel and sent the rate soaring, or the one on the right had run out of down travel and picked the inside wheel from the ground. Maybe both, but it happened all the more quickly in the face of that massive urge and it showed the limitations of transverse leaf springs.

But the Cobra was by no means the GTO's poor relation on the track. It didn't have the Ferrari's romantic side but in many ways it was probably a more effective racer for short dirty dogfights. Sure, when you have a clear track with long fast corners, you can use the Ferrari's balance to keep up the momentum, and where you have several miles flat out, like at Le Mans, the GTO's more beautiful body shape would count. But on a crowded track where you trip over slower cars and lose that corner entry speed, or somewhere with lots of slow corners, the Cobra's power would always come to the rescue. There was always more than the chassis could handle, so lose a bit to a mistake or have to lift because of traffic and you'd get it back with a prod of the boot. Former Ferrari world champion John Surtees (crotchety old cove that he is) recently admitted that he sometimes refused to drive a GTO because he knew he would trip over people while trying to use the car's strengths and he didn't need a shunt in a sports car while he was trying to win Grands Prix. Carroll Shelby, meanwhile, used the defeat at Le Mans in '63 to persuade Henry Ford to finance the Cobra Daytona Coupe. A sort of American GTO with long nose and tail, coil springs and the same big Ford engine which blew off Ferrari's GT cars at the 1964 Le Mans. The all-American crew-cut crew of Dan Gurney and Bob Bondurant won the class with fourth overall, just ahead of the ubiquitous Beurlys in a 1964-spec GTO.

These are both GT cars from a golden era where all the things you do for fun on a greasy roundabout were an essential part of every lap. Less a question of tyre compound, wing angle and ride height, more a question of putting the available ingredients together better than the next man. Inspiration in the cockpit could always make you faster. And it's physical. The steering is informative in a way modern racers can't imagine but it's heavy and you have to wedge yourself in different places to use it for different corners, use knees, elbows or whatever as braces because you are not belted to the point of pain in some deep Kevlar bucket. You work all the time and step out sweaty with sore feet and bruises you only find the following morning. You have to work at it, because 40 years down the road these are still massively fast cars in any terms.

A favourite then…? Well, that all depends whether you're asking as a romantic, a racer, or a businessman. Me, I just love driving them.

Thanks to Nick Mason, Ten Tenths, Nigel Hulme and everyone at Goodwood for their help with this feature.

Ferrari 250GTO

Engine	V12, 2953cc, 24v
Transmission	Five-speed manual, rwd
Max power	296bhp @ 8000rpm
Max torque	203lb ft @ 5500rpm
Weight (kerb)	1045kg
Power-to-weight	288bhp/ton
0-60mph	6.5sec
Max speed	150mph
Estimated value	£3,500,000-£4,000,000

AC Cobra

Engine	V8, 4726cc, 16v
Transmission	Four-speed manual, rwd
Max power	380bhp @ 6500rpm
Max torque	400lb ft @ 4500rpm
Weight (kerb)	1035kg
Power-to-weight	373bhp/ton
0-60mph	5.5sec
Max speed	160mph
Estimated value	£400,000-450,000